EXPLOSIONS IN THE PROCESS INDUSTRIES

Second Edition

A report of the Major Hazards Assessment Panel Overpressure Working Party

INSTITUTION OF CHEMICAL ENGINEERS

Published by
Institution of Chemical Engineers
Davis Building
165–189 Railway Terrace
Rugby, Warwickshire CV21 3HQ, UK

First Edition published as *The effects of explosions in the process industries (Overpressure Monograph)*

ISBN 0 85295 315 1

Printed in the United Kingdom by Galliard (Printers) Ltd, Great Yarmouth

PREFACE TO THE SECOND EDITION

The Major Hazards Assessment Panel was established in 1982 to provide a means for scientists and engineers, as professionals rather than as representatives of companies or other organizations, to comment on matters of current interest in the field of major chemical hazards. The full Panel, about sixty people, never meets. Instead a small Advisory Group selects subjects for study from among those suggested by the members of the Panel and Working Parties are set up to prepare reports on these subjects. The draft reports are sent to the full membership for comment.

The Institution of Chemical Engineers helped to set up the Panel and provides secretarial services and also publishes its reports. However the Panel is not responsible to the Institution and it is not a committee of the Institution. The authority of the Panel's reports is due solely to the professional reputation of its members, particularly the members of the Working Parties.

This Working Party, on overpressure, was established by the Major Hazards Assessment Panel and has as its terms of reference:

'to consider and report on the effects of industrial explosions at places to which the public have access (particularly houses and public roads) and the type and magnitude of the initiating events and how these may be calculated.'

For this Second Edition a second Working Party was appointed.

The new Working Party preferred to interpret its terms of reference broadly and to consider also some types of explosion which are significant to the process industry but, by themselves, are only marginally hazardous in the public domain. They took the view that such explosions may cause further damage to chemical plant which may give rise to explosions which could be significant to the public.

MEMBERSHIP OF THE WORKING PARTY

EDITOR
H. Phillips

WORKING PARTY

A. Cates	Shell
C.A. Catlin	British Gas
N. Edmondson	UKAEA Safety and Reliability Directorate
M. Goose	Health and Safety Executive
R. Merrifield	Health and Safety Executive
P.A. Moreton	AEA Technology, Non-Nuclear Projects Department
D.K. Pritchard	Health and Safety Executive
G.O. Thomas	University of Wales, Aberystwyth

Contributions from M. Mihsein, J. Moorhouse and R. J. Harris from British Gas are acknowledged with thanks, together with others who made valuable contribution through discussion.

The Working Party has rewritten the *Overpressure Monograph* to include recent advances in the state of knowledge on explosions and the methods used to predict their effects on buildings and other structures.

The views expressed in the monograph are not necessarily held by all contributors, nor do they necessarily represent the views of all their employers.

UNITS
Unless otherwise stated SI units are used. There are exceptions; for pressure 1 bar is approximately equal to 100,000 Pa; energy may be expressed as the equivalent mass of TNT, with 1 kg of TNT being equivalent to 4.6 MJ.

SYMBOLS
Symbols are re-defined in each section. The same symbol may be used with a different meaning in different sections.

SCALED UNITS
The axes of graphs may be labelled in non-dimensional units or scaled units. For non-dimensional units any self-consistent set of units may be used, such as SI units. Scaled units are not non–dimensional. The units to be used are stated and using other units will give incorrect results.

CONTENTS

1. INTRODUCTION

Each year the process industries handle a wide range of different materials and utilize them in many different types of chemical reactions. A significant proportion of the materials used are hazardous and many of the processes must be very carefully managed to ensure safe operation. Failure of either equipment or operational standards could, in some cases, result in an explosion which could take one of many forms. It could, for example, be the explosive failure of a pressure vessel due to overpressurization, or it could be a vapour cloud explosion (VCE) following an ignited release of flammable gas or vapour. That such events are rare is due to the attention paid to their prevention. Nevertheless it is necessary to be able to assess the probable consequences of an explosion, both in terms of damage to the plant itself and surrounding plants, and injury to the operating personnel and the general public. Such assessments can then inform judgements as to whether further safeguards are necessary and also assist in emergency planning.

Offshore safety is outside the scope of this Working Party, although many of the findings may be applicable offshore as well as onshore.

The principal legislation affecting onshore installations in Great Britain where there is a potential explosion hazard is the Health and Safety at Work etc Act 1974 (HSW). This enabling Act lays down, amongst other things, general requirements for safety of persons both on- and off-site. Specific requirements are given in Statutory Instruments (SI) made under the Act. Certain other Acts are made Relevant Statutory Provisions (RSP) to the 1974 Act.

There are many requirements in the legislation, to take steps aimed at preventing explosions. Although assessment of the hazards from explosions might be necessary to comply with the general requirements of the HSW Act, there is only one specific requirement for any assessment of explosion hazard to be carried out. This is set forth in the Control of Industrial Major Accident Hazard Regulations 1984, as amended (CIMAH). These regulations apply to sites having large quantities of hazardous substances. The actual amount varies from substance to substance, but the quantity is listed for each substance in Schedules 2 and 3 of the regulations.

The CIMAH Regulations implement the European Community's 'Seveso' Directive in Great Britain. The relevant parts of the SI are Regulation 7 and Schedule 6. Guidance has been published by the Health and Safety Executive, the enforcing authority for the Regulations[1]. The CIMAH Regulations require, according to circumstances, the preparation of a 'Safety Report', and on-site and off-site emergency plans. The Safety Report would include, amongst other items, hazard or risk assessments. Paragraphs 35 to 46 of the guide, on pages 63 to 65 give particular advice on the need for hazard assessment in particular cases.

The purpose of this monograph is to summarize in basic terms the present status of knowledge of explosions and to describe simple methods by which their effects in the process industries may be assessed. The monograph is not concerned with the manufacture of explosives, which is a highly specialized branch of the process industries, nor is it concerned with nuclear explosions. Nevertheless, the vast amount of research conducted on their properties, chiefly for military purposes, has many useful pointers to the accidental explosion which is the topic of this monograph.

Chapter 2 describes the types of explosion that might occur, and the way they might develop damaging overpressure and missiles. Chapter 3 describes the way in which overpressure and missiles might interact with structures and people to cause damage or injury. Chapter 4 describes methods for estimating the blast overpressure and missile velocity. Chapter 5 describes means of estimating the degree of injury and damage from the magnitude of the overpressure and missile velocity calculated in Chapter 4. Chapter 6 gives some worked examples of the calculations described in Chapters 4 and 5.

In view of the imprecise nature of the data available on potential accidents, it is considered that the simple methods reported here give an adequate assessment of the damage potential. However, more detailed methods are also available which might be appropriate for specific situations but are deemed beyond the scope of this text. Readers are advised to seek specialist advice before considering such methods, or where this text does not address their specific problem.

REFERENCE IN CHAPTER 1

1. HSE, 1984, *A Guide to the Control of Industrial Major Accident Hazard Regulations 1984, Booklet HS(R)21(rev)* (HMSO, UK).

2. EXPLOSIONS

An explosion may be defined as a process whereby a pressure wave is generated in air by a rapid release of energy. This definition encompasses widely differing events ranging from the trivial example of a spark discharge through sudden release of stored energy in a compressed gas to the extreme of chemical detonations and nuclear explosions.

Explosions can be listed according to the source of the energy. In the list below an attempt is made to order the explosions in decreasing damage potential, although there may be some overlap. As the reader proceeds through this report the characteristics of each type of explosion will be described with reference to the method for estimating its damage potential. Detailed discussion of the estimation methods follows later. Nuclear explosions are considered beyond the scope of this text.

2.1 HIGH EXPLOSIVES

Explosions of conventional high explosives are extremely rapid events known as detonations, and are characterized by the existence of a supersonic shock wave travelling in front of the reaction/flame front. As the shock front passes the boundary of the explosive charge it attenuates.

For detonation to propagate through the length of an explosive, the rate of energy loss behind the flame front must be no greater than some fraction of the rate of energy supply from the flame front. Reductions of these losses can be achieved by confining the explosive charge. Thus the amount of confinement has an effect on the nature of combustion/reaction. For example, ignition in a thin layer (few hundred grammes) of black powder spread thinly across a surface will produce a rapid burning. Ignition of this same quantity of material held inside a strong metal container can result in a violent explosion.

The United Nations scheme[1] for the classification of dangerous substances for transport recognizes the importance of confinement and prescribes tests for substances 'as packaged for transport'. By this scheme the explosives Class 1 is subdivided as shown in Table 2.1 (see page 4).

TABLE 2.1
Subdivisions of explosives Class 1

Hazard division (HD)	Definition
1.1	Substances and articles which present a mass explosion hazard.
1.2	Substances and articles which have a projection hazard but not a mass explosion hazard.
1.3	Substances and articles which have a fire hazard and either a minor blast hazard or a minor projection hazard or both, but not a mass explosion hazard.
1.4	Substances and articles which present no significant hazard.
1.5	Very insensitive substances which have a mass explosion hazard.
1.6	Extremely insensitive articles which do not have a mass explosion hazard.

In practice this could mean that a propellant packed in a particular way could well be classified as hazard division (HD) 1.3 — that is, on initiation it would exhibit thermal radiation effects only. This same propellant in a 'packaging' arrangement which affords greater confinement, however could well produce HD 1.1 type effects on initiation.

Accidental initiations of high explosives can and have resulted via a number of mechanisms, viz; fire exposure, excessive impact and/or friction and explosives instability (for example, due to ageing).

2.2 ENERGETIC SUBSTANCES

Energetic substances can be considered to be substances which are not formally regarded as explosives (that is, not assigned to Class 1 of the UN scheme) and yet have sufficient energy available to decompose exothermally, to deflagrate or, in certain circumstances, to detonate.

Examples of substances currently listed in the UN scheme which are not transported under the provisions of Class 1, but which have shown either explosive behaviour in incidents or in laboratory testing include nitromethane, *n*-propyl and iso-propyl nitrate, ammonium nitrate, ammonium perchlorate and *m*-dinitrobenzene. Many other substances outside Class 1 (for example, organic peroxides) are diluted or packaged in such a way as to reduce the level of confinement and hence explosibility. In an industrial situation these same materials could be subsequently transferred to a situation where the confinement is greater; this could possibly produce significant explosion effects on initiation.

Detonations of energetic substances are usually 'lower order' (less violent) events than detonation of high explosives, releasing their energy over a longer, although still short, period of time.

A useful comparison of the explosion hazards associated with explosives and industrial chemicals with explosive properties is given by Merrifield and Roberts [2].

2.3 GAS AND VAPOUR CLOUD EXPLOSIONS

'Vapour cloud explosion' (VCE) is the term used to describe the explosive combustion of the gas or vapour cloud formed by the release of flammable material, referred to here as the fuel, into the atmosphere. The vapour cloud can result from both gas-phase releases and the evaporation of liquid fuels.

Because liquefied vapours have a much higher density than compressed gases, the formation of clouds containing tonnes of vapour is much more likely than the formation of clouds containing tonnes of gas.

In the following sections those factors which affect explosion severity are described. However, it is important to appreciate that blast damage will not always result from a release of flammable material. For example, a vapour cloud on open, unobstructed terrain is unlikely to produce damaging overpressure in any but the most exceptional circumstances (for example, a gas which is substantially more reactive than typical hydrocarbon fuels or there is a very high energy ignition). The following sections give methods for estimating the worst-case overpressure should an explosion occur.

GAS AND VAPOUR CLOUD FORMATION

The rate of energy release, which is proportional to the rate of combustion, depends upon the mixture composition, and is a maximum at a point close to the stoichiometric air-fuel ratio. The explosion overpressure will therefore also be highest under these conditions.

There are lower limits to the concentrations of fuel and air in the vapour below which the vapour will not burn. In some instances, therefore, a detailed consideration of how the fuel may disperse may discount the possibility of a vapour cloud explosion by showing that a flammable mixture cannot form. Also at mixture concentrations above the upper limit of flammability the rate of burning is controlled by the rate of mixing with air, and therefore it is not likely that destructive blast will be generated by rich mixtures. However, predicting how a gas will disperse is a complex science which is outside the remit of this text. Such a study must consider factors such as whether the gas is lighter or heavier than air, the release rate and direction, atmospheric conditions, the presence of obstructions or confinement. When there is any doubt about the fuel-air concentration in a vapour cloud, conservative predictions of overpressure can be obtained by assuming that the fuel and air have mixed in stoichiometric proportion.

IGNITION

Every effort should be made to eliminate potential ignition sources, for example by the use of explosion protected electrical equipment, area classification, permit to work procedures. However, even with such precautions it is not wise to assume that a possible ignition source does not exist.

EXPLOSION EFFICIENCY

The magnitude of the blast wave is determined principally by the rate of combustion, and as the density of the unburnt mixture decreases by a factor of approximately eight as it burns (at constant pressure), this can be related to the rate at which the products expand, or equivalently, the rate at which an increasing volume of product gas is generated. As the products expand, and because of the inertia of the surrounding atmosphere, higher blast pressures are generated for higher combustion rates. This in turn increases the efficiency with which the chemical energy of the vapour is converted to the mechanical energy of the blast wave. Thus one of the key parameters determining the efficiency of the explosion is the maximum flame speed, or equivalently the peak overpressure at the flame. The other factor influencing the far-field overpressure is the physical size of that part of the cloud which is consumed at a high combustion rate. With a very low flame speed, for example in an unconfined and unobstructed vapour cloud, the overpressure is generally too low to cause blast damage.

CONFINEMENT

When solid structures enclose parts of the vapour cloud the explosion is said to be 'confined'. This includes combustion inside buildings, equipment housings and process equipment. With insufficient gaps in the confinement to allow the expanding gases to escape, or 'vent', the pressure inside the confined region will rise above that in the external atmosphere and place the confining structure under a load that could rupture it should it be insufficiently strong. A common example is when the explosion occurs in a building, when windows and non load-bearing walls can fail. Projectiles in the form of high-velocity glass fragments and falling debris can then pose significant hazards to people.

Hazards are posed by the venting gas which can travel large distances from the confined volume. The velocity of the venting gas can be very high, even for moderate overpressures inside the confined region, which imposes high side loading on structures within the gas jet. Such loads could cause equipment to collapse and generate high-velocity projectiles. When the burned gas is vented, the combustion of flammable gas already vented and ignited by the burning gas jet can be very rapid. Such an 'external explosion' can provide a significant 'source' of blast, especially if there are obstacles within the jet, when high flame speeds and overpressures can be generated and sustained. The external explosion, by restricting venting, can cause the pressure within the confinement to rise to higher levels than had the external explosion not occurred.

The 'external explosion' is significantly enhanced if initially there is a flammable cloud external to the vessel. The jet of flame is a very powerful source of ignition for the external cloud.

Damaging overpressures can be circumvented in many instances, however, if the vent area is made large enough. Mitigating damaging overpressures in this way is known as explosion venting for which there are a number of simple guidelines [3,4,5,6,7,8].

OBSTRUCTIONS/PARTIAL CONFINEMENT

It is accepted that repeated obstructions are a key factor in generating high flame speeds in an explosion and thereby high explosion efficiencies. Such obstructions are common in industrial processes where large volumes can be occupied by process vessels and interconnecting pipe work. This type of plant is otherwise open to the atmosphere and consequently vapour clouds can be either formed within or drift into such areas. An alternative and commonly used term to describe explosions in such situations is a 'partially confined explosion'.

A flame can accelerate rapidly if it repeatedly meets obstacles. However, the flame will only maintain high speeds if it continues to meet obstructions. Once the flame has passed the boundary of such an obstructed region, or the congestion level lessens, the flame will start to decelerate.

For a given fraction of the volume occupied by obstructions (volume blockage) the potential for flame acceleration increases as the diameter of the obstructions gets smaller. For a given diameter of obstruction the flame acceleration potential increases with increasing volume blockage.

Particular concern should be shown for any situation in which there are both repeated obstructions and confinement. This may arise, for example, when the obstructions are confined below a ceiling, or when the route for vented gas lies between two walls. Also an enclosure may contain repeated obstructions, giving high overpressure which is difficult, if not impossible, to relieve by venting.

There is no simple guideline for predicting flame acceleration and expert advice should be sought should an accurate assessment be required. Flames will not carry on accelerating indefinitely, however, and will ultimately reach a maximum flame speed. A conservative estimate of explosion overpressure and consequences can therefore be obtained by assuming an appropriate maximum flame speed.

The maximum speed achievable depends on the fuel type, the obstacles and confinement. Such speeds can in fact be very high if the explosion has a suitably energetic ignition. The absolute limit to the explosion efficiency occurs when the vapour cloud detonates (see page 9). Detonative combustion is a particular type of flame which, once initiated, maintains its speed without needing obstacles.

In the absence of detonation, once the flame leaves the obstacles, the flame speed can be assumed to fall rapidly. Consequently if there are adequately large gaps between regions containing obstacles then a high speed flame will not be able to transmit from one region to another. In these situations the volume of vapour that can give rise to blast is related to the volume of the congested area and not to that of the total vapour cloud (assuming the cloud has totally engulfed the congested area). There is a delay between the blasts produced by each obstructed region, so that, at any point, the maximum overpressure is the maximum of the overpressures produced by each obstructed volume.

DETONATION

Detonative combustion is sustained by a shock wave which propagates at the leading edge of the combustion front. The shock has a pressure of typically 20 atmosphere. The propagation speed is consequently very fast, being close to 2 kilometres per second for a stoichiometric hydrocarbon-air mixture.

Once initiated, detonation waves pose a major hazard, not only because they are very efficient at generating blast but also because their propagation is principally restricted by the concentration of the fuel in the vapour cloud. The combustion front requires neither obstruction nor turbulence to maintain its speed.

The transition to detonation for most materials is thought to be unlikely in explosions in chemical plant. In this context, the Advisory Committee on Major Hazards[9] has recommended that the possibility of detonation be disregarded. It is suggested that this view may be too prescriptive for more reactive materials such as ethylene oxide and ethylene. Such materials have been observed to make a transition to detonation whilst the flame is accelerating through obstacles and hence, for the more reactive materials, the possibility of a detonation in a highly confined layout needs to be considered. For further information see Nettleton [10].

KEY CONSIDERATIONS

- Assess whether the explosion venting is adequate for areas of confinement.
- Minimize the possibility of gas being vented by a confined explosion into a congested region.
- Minimize the likelihood of vapour cloud forming in heavily congested region.
- Minimize extensive regions of congestion; if there are adequate gaps the flame acceleration can be curbed.
- Minimize confinement around congested areas.
- Minimize material that might become a projectile.
- Inventories of easily detonable flammable vapour should be minimized whenever possible.
- All steps should be taken to avoid flammable concentrations of easily detonable material forming in regions of congestion and/or confinement.

2.4 DUST EXPLOSIONS

Dust explosions are a hazard to plant, but the resulting overpressure does not generally propagate far outside the plant. The effect of a dust explosion is to rupture the ventilation ducts and buildings containing the explosion. Since these structures generally fail at a low overpressure the distance over which a blast wave can create further damage is very small. Following rupture of the container or building, a dust cloud might be expelled into the atmosphere. The factors influencing the violence of the explosion in the atmosphere are similar to those for the gaseous explosion, but generally the extent of the dust cloud is smaller. A typical site for a dust explosion is a grain silo. Dust explosions are not considered further, but the reader is directed to the IChemE dust explosion prevention and protection guides[12,13,14]. A more serious hazard than overpressure is the vented fireball that may emerge from a vented explosion. Fireballs may be much larger than the vessel in which the explosion started and jet flames exceeding 23 m in length have been recorded.

2.5 BURSTING PRESSURE VESSELS

The sudden release of energy from rupture of a pressure vessel can give rise to damaging pressure waves and high velocity fragments.

Pressure bursts may be the consequence of sudden pressurization of the vessel following a runaway reaction or may be the consequence of excessive heating of the contained fluid should the vessel be surrounded by fire. If the vessel contained compressed gases the expanding pressure wave is a consequence of the expansion of those gases. If the vessel contained a hot liquid, or a liquefied gas at ambient temperature, however, rapid vaporization of the liquid following rupture of the vessel will add to the overpressure in the expanding pressure wave.

Overpressurization, leading to fracture of the vessel, has been known to occur when a vessel contains stable layers of hot and cold liquids. Should the layers become unstable, or be stirred, the rapid mixing of a hot liquid with a cold liquid might cause vaporization of the low temperature liquid and rapid pressurization of the vessel.

If the vessel contains a flammable, hot liquid the combustion of the liquid or its vapour in contact with air can augment the pressure wave, although in most instances the thermal radiation from the resulting fireball is the main hazard.

A vessel containing a liquid (with no air space) is not likely to produce an expanding wave of overpressure if there is no evaporation or combustion following rupture of the vessel.

2.6 AEROSOL EXPLOSIONS

Liquid aerosols are used extensively and quite safely, when properly controlled, for applications such as paint spraying. Aerosols can also be formed accidentally when a pressurized or flashing liquid release occurs from a processing or storage situation. The aerosol produced from such an accidental release may subsequently rain out, or be transported downwind in a hybrid aerosol/vapour mixture. The possibility of a subsequent fire or explosion of such a mixture is a topic which has had little attention. The work that has been carried out has generally done little to characterize the aerosols in terms of nature and time dependency. Another important factor is the amount of air entrainment. No theoretical model exists to quantify this entrainment or its dependence on the release conditions or the properties of the material released. The work of Kopyt *et al*[14] on aerosol clouds (8.5 m radius; 500 m^3) of fuel oil, etc indicated flame velocities approaching 100 m/s. These high velocities were thought to be due to a combination of turbulence, non-uniform fuel mixture density and a pre-heating of droplets (by thermal radiation) ahead of the flame front.

2.7 RAPID PHASE TRANSITION (RPT)

THE GENERAL MECHANISM OF AN RPT

When two liquids with significantly different temperatures are brought into contact there is a possibility that a rapid phase transition (RPT) might occur. This can arise when the hot liquid or hot particle is capable of heating the cooler liquid to above its boiling point, that is, up to or near its superheat limit temperature*. The rapid phase change is accompanied by a large volumetric

---

* A liquid can be heated to above its normal boiling point, thereby becoming superheated. This can only occur up to a certain temperature, the superheat limit temperature, at which the liquid boils spontaneously.

expansion which occurs so rapidly that a shock wave is produced. Under some circumstances RPTs have caused structural damage and loss of life.

RPTs are known to have occurred in metal manufacturing and ore extraction upon contact of molten metal with water and in paper making processes when molten salts are quenched in water. In the nuclear industry research has been done to evaluate the consequences of an accidental contact of molten sodium or other molten metals with water, although with sodium and water chemical reaction can also add to the damaging effect. The potential for RPTs occurring also arises in accidents involving the handling of gases liquefied by cooling, should they be spilled onto a warmer fluid, typically water.

A summary of the known fluids and situations for which RPTs have been known to occur in industrial situations or experimentally is given in Table 2.2. Inclusion in the list does not imply that an RPT will always occur.

An RPT can generate blast waves similar to those produced by high explosives except that the maximum energy yield is limited by the superheat energy of the cold fluid involved. In fact, only a small fraction of the available superheat energy is transferred into the shock wave so in energy terms RPTs are less efficient than high explosive at generating blast.

PREVENTION/MITIGATION OF RPTS

No generalized method is currently available for predicting the consequences of an RPT nor for ensuring that RPTs do not occur, and this is likely to remain the case due to the wide variety of fluids and situations in which RPTs can arise. The only fully safe approach is to avoid contact between fluids with significantly different temperatures. The quantity of the fluids involved and the degree of confinement should be minimized wherever possible.

Means of preventing RPTs have arisen directly out of industrial incidents. In the metal casting industry, the occurrence of an RPT between molten metal and water has led to the development of a variety of coatings for the water container which modify the manner of contact and thereby prevent the occurrence of RPTs. In the paper industry water and steam jets have been used to fragment and cool the hot smelt prior to entering the quench bath, which has also been found to be an effective means of prevention. In addition the presence of soluble salts in the quench water has been found to influence the occurrence of RPTs and studies with various salts have identified those which promote and prevent RPTs.

TABLE 2.2
Summary of systems in which rapid phase transitions have been observed

Cold fluid	Hot fluid	Notes
Water	Molten metals including aluminium, steel, nickel, magnesium, tin, titanium, tungsten, molybdenum, copper.	Industrial incidents mainly with aluminium. Explosion probability dependent on degree of pre-fragmentation of molten metal and also surface coatings of water vessel.
Water	Smelt-molten salt mixture containing primarily sodium carbonate — product of paper pulp industry.	Severity dependent on minor constituents of salt mixture and solutes in the water.
Cryogens, refrigerants including hydrocarbons, C_1–C_4 hydrocarbons and hydrocarbon mixtures	Water and other non-freezing liquid substrates	Mainly observed in laboratory tests and field trials investigating RPTs.
Water	Hot oil	When hot oil contacts a water layer in a storage tank or plant equipment the water is vaporized with explosive violence. The tank or equipment may rupture and a mixture of oil and steam may be expelled.

RECOMMENDATIONS

- Avoid opportunities for liquids of significantly different temperature to come into contact with one another.
- Minimize the inventory of the fluids.
- Avoid such an interaction occurring in confined regions.

REFERENCES IN CHAPTER 2

1. *Recommendations on the Transport of Dangerous Goods*, Seventh revised edition 1991 (United Nations, New York, USA).
2. Merrifield, R. and Roberts, T.A., 1991, A comparison of the explosion hazards associated with the transport of explosives and industrial chemicals with explosive properties, *IChemE Symposium Series No 124, Hazards XI: New Directions in Process Safety* (IChemE, UK).
3. Chamberlain, G.A., September 1989, *The nature and mitigation of vapour cloud explosions*, A paper prepared for the Cullen Inquiry into the Piper Alpha Incident.
4. Harris, R.J., 1983, *Gas Explosions in Buildings and Heating Plant* (E. and F.N. Spon, London, UK).
5. Health and Safety Executive, 1987, *Health and Safety at Work Booklet, HS(G)16* (HMSO, UK).
6. Health and Safety Executive, 1980, *Health and Safety Series Booklet HS(G)11* (HMSO, UK).
7. Lunn, G. A., 1984, *Venting Gas and Dust Explosions — A Review* (IChemE, UK).
8. *NFPA 68, Guide for Venting Deflagrations*, 1988 (National Fire Protection Association, USA).
9. Health and Safety Commission, 1979, *Advisory Committee on Major Hazards*, Second Report, (HMSO, UK).
10. Nettleton, M.A., 1987, *Gaseous Detonations, their Nature, Effects and Control* (Chapman and Hall, London, UK).
11. Lunn, G.A., 1992, *Dust Explosion Prevention and Protection Guide: Part 1, Venting*, 2nd Edition (IChemE, UK).
12. Schofield, C. and Abbott, J. A., 1988, Ibid, *Part 2, Ignition Prevention, Containment, Inerting, Suppression and Isolation* (IChemE, UK).
13. Lunn, G.A., 1988, Ibid, Part 3, *The Venting of Weak Explosions and the Effect of Vent Ducts* (IChemE, UK).
14. Kopyt, N.Kh., Struchaev, A.I., Krasnoshchokov, Yu.I., Rogov, N.K. and Shamshov, K.N., 1989, *Fizika Gorienya i Vzryva*, 25: 21–28.

FURTHER READING

1. Bankoff, S.G., 1989, Vapor explosions: a critical review, *Proc 6th Int Heat Transfer Conference*, 6: 355–360.
2. Baker, W.C., Cox, P.A., Westine, P.S., Kulesz, J.J. and Strehlow, R.A., 1983, *Explosion Hazards and Evaluation* (Elsevier, Amsterdam, The Netherlands).
3. Bjorkhaug, M., October 1986, Flame acceleration in obstructed radial geometries, *PhD Thesis* (Department of Chemistry, City University, London, UK).

4. Board, S.J., Hall, R.W. and Hall R.S., March 1975, Detonation of fuel — coolant explosions, *Nature*, 254 (5489): 319–321.
5. Eckhoff, R.K., 1991, *Dust Explosions in the Process Industries* (Butterworth Heinemann).
6. Enger, T. and Hartman, D.E., 1972, Rapid phase transformations during LNG spillage on water, *LNG–3 Conference* (Washington).
7. Epstein, S.G., Loviza, L.E. and Gommi, J.V., July 1988, *Proc Workshop on Causes and Prevention of Melt-Water Interactions, Sandia National Laboratories, Albuquerque, New Mexico.*
8. Harris, R.J. and Wickens, M.J., 1989, Understanding of vapour cloud explosions — an experimental study, *55th Autumn Meeting, Communication 1408* (Institute of Gas Engineers, UK).
9. Harrison, A.J. and Eyre, J.A., 1987, External explosions as a result of explosion venting, *Combustion Science and Technology*, 52: 91–106.
10. Harrison, A.J. and Eyre, J.A., September 1986, Vapour cloud explosions — the effect of obstacles and jet ignition on the combustion of gas clouds, *Proceedings of 5th Symposium Intl of Loss Prevention and Safety Promotion in the Process Industries, Cannes, France.*
11. Hjertager, B.H., Fuhre, K. and Bjorkhaug, M., October 1988, Gas explosion experiments in 1:33 and 1:5 scale offshore separator and compressor modules using stoichiometric homogeneous fuel/air clouds, *J Loss Prev Process Ind,* 1.
12. Lee, J.H.S. and Moen, I.O., 1980, The mechanism of transition from deflagration to detonation in vapour cloud explosions, *Progress in Energy and Combustion Science*, 6: 359–389.
13. Reid, R.C., 1983, Rapid phase transitions from liquid to vapor, *Adv Chem Eng,* 12: 105–208.
14. Van Wingerden, C.J.M. and Zeeuwen, J.P., 1986, Investigation of the explosion enhancing properties of a pipe-rack-like obstacle array, *Progress in Astronautics and Aeronautics*, 106: 53–65, (AIAA, New York, USA).
15. Van Wingerden, C.J.M., 1989, On the scaling of vapour cloud explosion experiments, *Chemical Engineering Research and Design,* 67: 339–347.

3. INJURIES AND DAMAGE

3.1 INTRODUCTION

An explosion has already been defined as a process whereby a pressure wave is generated in air by a rapid release of energy. It is this pressure wave, often referred to as the blast wave, that is largely responsible for the injuries and damage caused.

The blast wave produced by an explosion interacts with any objects in its path, imposing a mechanical load on them. The blast damage results from the response to this loading — for example, deformation of structures, people being knocked over.

People may be injured and structures damaged by explosions through several other mechanisms — missile strike, crater formation, ground shock and fire. Ground shock and cratering resulting from condensed phase explosions or detonations can be important considerations for buried targets such as pipelines and hence cause 'domino' or 'knock-on' effects, but for targets above ground the effects are normally small compared with those of the blast wave. Accordingly, cratering and ground shock have not been considered further in this monograph. Missiles can be generated by condensed phase explosions, pressure vessel bursts and confined or partially confined explosions.

3.2 THE BLAST WAVE

Before considering how blast waves interact with structures and objects, the parameters required to characterize blast waves need to be defined. This is best done by reference to an ideal blast wave — that is, the blast produced by an instantaneous release of energy at a point source. The pressure versus time history for an ideal blast wave is shown in Figure 3.1. On arrival of the blast wave there is an instant rise in pressure, the shock front, followed by a fairly rapid decay to a pressure below atmospheric pressure, before the pressure returns to atmospheric. It is usual to refer to the pressure produced by a blast wave in terms of the overpressure — that is, the difference between the absolute blast pressure and atmospheric pressure. Thus the overpressure can be positive

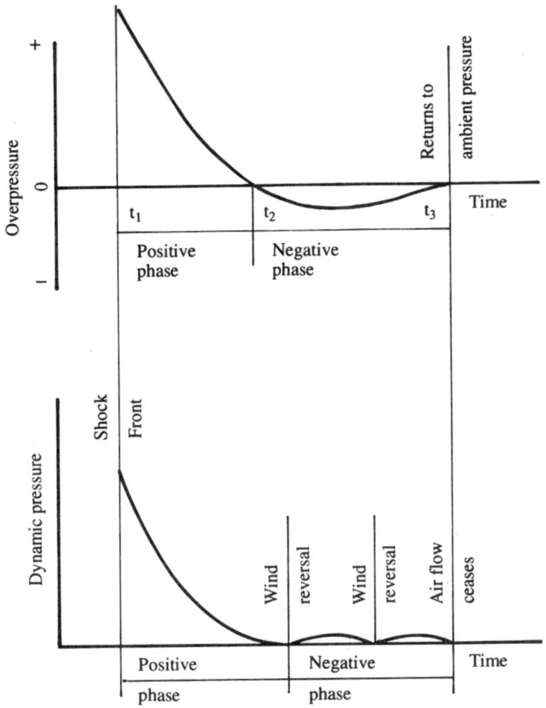

Figure 3.1 Variation of overpressure and dynamic pressure with time, at a fixed location, for an ideal blast wave.

or negative, giving positive and negative phases. The other parameters often quoted are the peak value (maximum overpressure attained), the duration and the impulse. The duration is the time for which the overpressure persists; for example, from Figure 3.1 the positive duration is $t_2 - t_1$ and the negative duration is $t_3 - t_2$. The impulse is the area under the overpressure versus time curve and again it is usual to give a value for the positive and negative phases.

To estimate the injuries and damage caused by a blast wave the dynamic pressures are also required. The dynamic pressure results from the air movement induced by the blast wave, often referred to as the ‘blast wind’, and is proportional to the square of the air velocity and to the density of the air behind the blast wave. Figure 3.1 also shows the dynamic pressure/time history for an ideal blast wave. Like the overpressure it is characterized by a peak value, duration and impulse.

In practice, no explosion produces an ideal blast wave. Blast waves from high explosives, like TNT, closely approach the ideal wave, because of the relatively small volume of the explosive material and the rapid rate of energy release. This is not the case for vapour cloud explosions, as the volume of vapour/air mixture is large and the rate of energy release is relatively slow. Thus a vapour cloud blast wave has a lower peak overpressure and a longer duration than a blast wave from a high explosive of the same total energy release. The 'shape' of the pressure/time profile also differs from the ideal wave. The leading edge is not always a shock wave (that is, not an instantaneous pressure rise), in which case a rise time is also required to characterize the blast wave fully, and the negative phase, unlike the ideal wave, can be appreciable. A large negative phase may be significant when it comes to assessing blast damage; for example, broken windows may be sucked out of a building.

Scaling laws have been established for explosions producing near ideal blast waves in geometrically similar systems, which allow the blast wave properties for an explosion to be extrapolated from results obtained under different conditions. For example, on plotting overpressure divided by atmospheric pressure against distance divided by the cube root of the mass of the explosive, all points fall on the same curve (for a given explosive)[1]. The

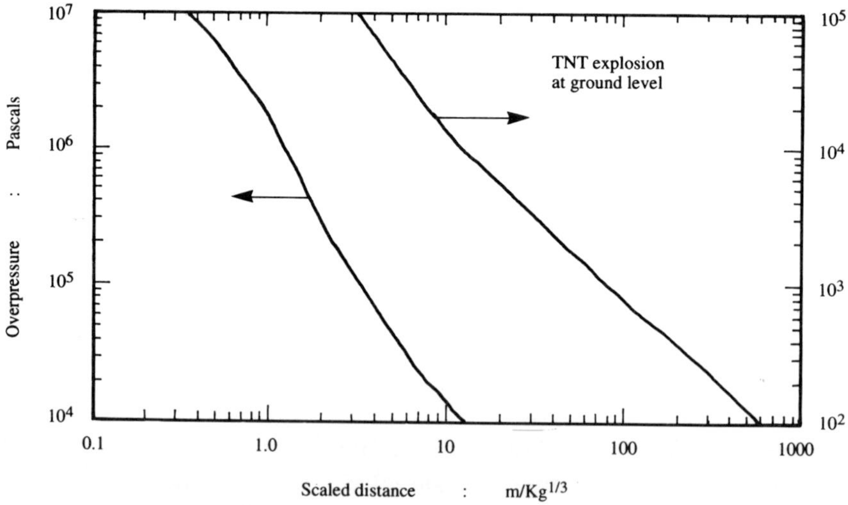

Figure 3.2 Side-on explosion pressure for a TNT explosion at ground level. Distance scaled by dividing by cube root of the mass of TNT. (Baker *et al*[2].) (Reproduced by permission of Elsevier, Amsterdam, The Netherlands.)

curve for the blast from a hemispherical cloud at ground level, with ignition at its centre — that is, at ground level — is reproduced as Figure 3.2, while the corresponding curve for impulse (which is also scaled like distance) is Figure 3.3. The curves can be used to predict the overpressure and impulse produced by different masses of explosive at any distance from the explosion centre and at any altitude (atmospheric pressure). At any scaled distance the overpressure remains the same, but impulse is larger as the explosion energy release becomes larger. The other blast wave parameters can also be scaled[2,3].

For a spherical cloud, with ignition at its centre, the blast overpressure and scaled impulse are obtained with double the mass in the scaled distances of Figures 3.2 and 3.3. Such a spherical cloud might be the result of failure of a pressure vessel in an elevated structure.

The loading, or force, on a structure or object due to the impingement of a blast wave on it, can either be as a result of diffraction loading or drag loading[2,3].

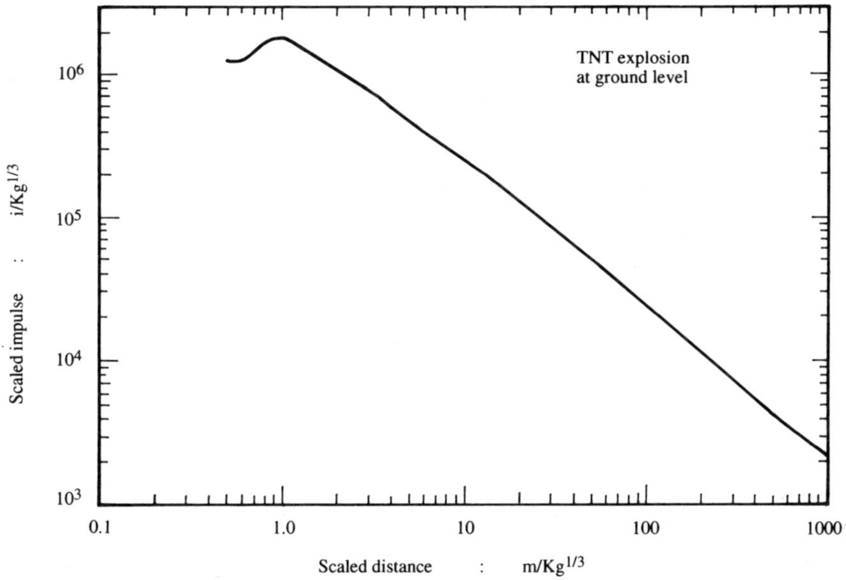

Figure 3.3 Scaled impulse for a TNT explosion at ground level. Impulse = positive overpressure times its duration, impulse and distance scaled by dividing by the cube root of the mass of TNT. (Baker *et al* [2].) (Reproduced by permission of Elsevier, Amsterdam, The Netherlands.)

Diffraction loading results from the pressure difference across the structure or object as the blast wave passes over it, and can be illustrated by considering an ideal blast wave interacting with a closed box-like structure. When the blast wave strikes the front face it is reflected, leading to an instantaneous pressure acting on the front face that is greater than the peak pressure of the incident blast wave. As the blast wave continues to move forward, the overpressure on the front face drops rapidly to the stagnation pressure — that is, the overpressure produced by the blast wave without reflection plus an added pressure due to the dynamic (blast wind) pressure. At the same time, the blast wave begins to act on the top and sides of the structure, and when the blast wave reaches the back end of the structure the pressure also begins to increase on the rear face. Once the structure is completely engulfed by the blast wave, approximately equal pressures will be acting on all faces of the structure. However, a small pressure difference will persist between the front and rear faces throughout the positive phase of the blast, since the rear face is shielded from the blast wind. The net horizontal load acting on the structure will thus be due to the pressure difference between the front and rear faces.

Thin structures such as pipes and girders are rapidly engulfed by the blast wave and the overpressure acting on each side equalizes. In these cases the loading is almost entirely due to the dynamic pressure acting on the structure and is called drag loading. The drag load is calculated by multiplying the dynamic pressure (a function of density and velocity) by the projected area normal to the blast wave and the drag coefficient.

3.3 EFFECTS ON STRUCTURES AND PLANT

The extent to which a structure is deformed by a blast wave will depend on the mechanical properties of the material — for example, yield point and ultimate tensile strength.

The damage to an item also depends on how quickly it responds to the blast loading. The natural period of vibration of the structure or object is usually used as a measure of the response. Thus small, flimsy structures will respond more quickly to a blast wave than large, rigid structures. If the duration of the blast wave is long compared to the natural period of vibration, the loading will be essentially static and the peak load will determine the degree of blast damage. For durations less than the natural period of vibration, the blast damage will depend on the impulse.

The same degree of damage to a structure can be caused by different combinations of peak overpressure and impulse. This has given rise to the concept of the pressure-impulse or PI diagram, where combinations of overpressure and impulse giving the same degree of damage (isodamage curves) are plotted on axes of pressure versus impulse. The PI diagram shown in Figure 3.4 is for a simple single degree of freedom elastic structure loaded with an ideal blast wave. There are three distinct loading regimes, two of which (the impulsive and static or quasi-static) have already been described. In the third regime, damage depends on both the overpressure and impulse. Further information on PI diagrams for more complex structures and how the incorporation of non-ideal waves and material plasticity affect the shape of the isodamage curves can be found in Baker *et al*[2].

The message from this discussion is that the extent of blast damage in many cases does not just depend on the peak overpressure. To predict blast damage accurately requires information on the other blast wave parameters, like impulse and duration, and also information on the mechanical properties of the

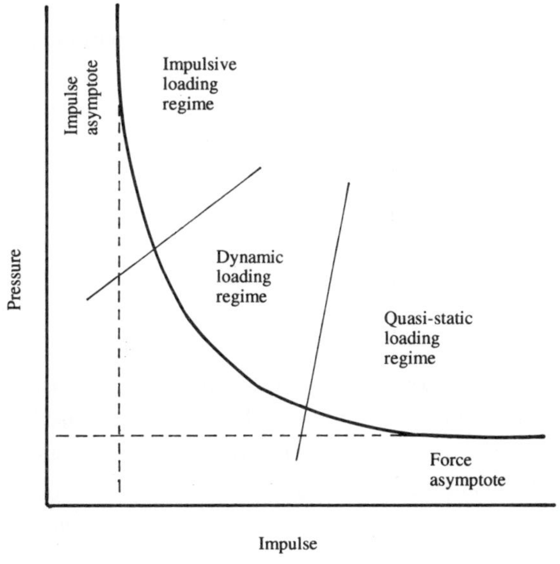

Figure 3.4 PI diagram for single degree of freedom elastic structure loaded with an ideal blast wave.

structure or object. For a given overpressure the impulse from a vapour cloud explosion is greater than for an ideal explosion, and that from a large explosion is greater than from a smaller explosion.

However, for the purpose of making a preliminary assessment of the possible off-site damage from an accident, it has been found adequate to relate the damage to maximum overpressure. This simplification is most valid for brittle targets such as brick or concrete structures where the effect of impulse is much less pronounced.

3.4 EFFECTS ON PEOPLE

Death and injuries to people from explosions can result from direct (primary) or indirect (secondary and tertiary) effects. Primary effects involve the direct effects of blast upon sensitive human organs such as the ears and lungs. Compared with indirect effects, considerable overpressures are required for fatalities to occur, and consequently people need to be fairly close to the scene of the explosion. Secondary effects are associated with building collapse or the impact of debris and fragments from damaged building structures and the vehicle or container in which the explosives are held. Predictive injury and fatality models for fragment/debris from high explosives are particularly difficult and outside the scope of this monograph. Tertiary blast injuries may occur with whole body impacts, when people are displaced or swept away, or due to the violent movement of internal organs within the body. For people outdoors tertiary effects are dominant.

3.5 MISSILES

Significant damage in many accidental explosions can be caused by the impact of fragments or objects which are generated during the explosion. The fragments, missiles or projectiles are considered a potential hazard because of their capability to escalate an incident. In this monograph the damage caused by missiles from accidental industrial explosions is considered. Missiles due to condensed explosives are not considered.

There are two possible types of projectiles/missiles. Primary missiles are generated due to the rupture of pressurized equipment, such as pressure vessels, or from the failure of rotating machinery such as gas turbines and pumps. The number of missiles generated by the rupture of pressurized equipment or the failure of rotating machinery, and their energy, have been reported in various publications (Baum[4], Brown and Smith[5], Chicken[6], Holder[7], Baker *et al*[8]). The energies of a primary missile will depend on its mass which in turn depends on the mode of failure of the system.

The other type of a projectile/missile generated during an explosion is a secondary one. These are generated from the interaction of blast waves with either loose (unconstrained) objects or objects broken off during the explosion (constrained). These objects can be accelerated to velocities which can cause impact damage. The objects could be small tools, loose debris and other structures disrupted by the explosion.

REFERENCES IN CHAPTER 3

1. Hopkinson, B., 1915, *British Ordnance Board Minutes 13565.*
2. Glasstone, S. and Dolan, P.J., 1977, *The Effects of Nuclear Weapons* (Third Edition) (United States Department of Defense).
3. Baker, W.D., Cox, P.A., Westine, P.S., Kulesz, J.J. and Strehlow, R.A., 1983, *Explosion Hazards and Evaluation* (Elsevier, Amsterdam, The Netherlands).
4. Baum, M.R., 1987, Disruptive failure of pressure vessels: preliminary design guidelines for fragment velocity and the extent of the hazard zone, *Recent Advances in Structural Dynamics*, 124: 175–181 (ASME Publication PVP).
5. Brown, A.L. and Smith, J.B., 1944, *Mechanical Engineering (US)*, 66: 392.
6. Chicken, J.C., 1980, Discussion at BNES Symposium, *J BNES*, 19 (3).
7. Holden, P. L., 1986, *Assessment of Missile Hazard: Review of Incident Experience Relevant to Major Hazard Plant, HSE/SRD/PD025 – WP2.*
8. Baker, W.E., Cox, P.A., Westine, P.S., Kulessz, J.J. and Strehlow, R.A., 1983, *Explosion Hazards and Evaluation* (Elsevier, Amsterdam, The Netherlands).

4. ESTIMATION OF BLAST AND MISSILE GENERATION

4.1 INTRODUCTION

Table 4.1 may assist in directing the reader to appropriate methods for estimation of blast and missile generation in particular cases.

For high explosives, energetic substances and phase change explosions the TNT equivalence method is the only one that can be recommended.

For gas and vapour explosions, specific methods exist, in addition to the TNT method which is based on the total mass of flammable material released. All the specific methods have one feature in common: they treat separately those parts of the gas or vapour cloud in confined or obstructed regions, and those parts that are unconfined.

TABLE 4.1
Methods for estimation of explosion overpressure and impulse

Explosion source	Assessment methods		
	TNT equivalence method 4.2	Pressure vessel method 4.7	Vapour cloud methods 4.3, 4, 5, 6
High explosives	Applicable	Not applicable	Not applicable
Energetic substances	Applicable	Not applicable	Not applicable
Bursting pressure vessels	Applicable	Applicable	Not applicable
Gas and vapour explosions	Applicable	Not applicable	Applicable
Dust and mist explosions	Applicable	Not applicable	Not applicable
Rapid phase transitions	Applicable	Not applicable	Not applicable

For bursting pressure vessels the TNT equivalence method may be used but there are also other methods that might be used, taking into account the differences between the pressure profile from a TNT explosion and the blast wave from an vessel rupture.

4.2 THE TNT EQUIVALENCE METHOD

INTRODUCTION

The basis of the TNT equivalence method for the modelling of the effects of explosions is that the energy being released is equated to a mass of TNT which, on explosion, would give an equivalent amount of damage. Once an equivalent mass of TNT has been assumed the variation of overpressure and impulse may be determined by reference to graphs or equations. These relationships are based on data from accidental and deliberate explosions of TNT and similar substances.

Equivalent mass of TNT = mass of explosive (or fuel) × equivalence factor

where:

- equivalence factor = energy ratio factor × efficiency factor;
- energy ratio factor is the explosion energy per unit mass of explosive (or fuel) compared with the explosion energy in a unit mass of TNT;
- efficiency factor is the proportion of the available combustion energy that is released in the explosion.

For some explosions the energy release may be calculated directly, in which case:

$$\text{TNT equivalence} = \frac{\text{explosion energy}}{4.6\ \text{MJ kg}^{-1}} \times \text{efficiency factor}$$

where 4.6 MJ kg^{-1} is the energy content of TNT.

Having found the equivalent mass of TNT the blast overpressure and impulse can be found from data obtained from the experimental explosions of a TNT charge placed on the ground (see Figures 3.2 and 3.3, page 19). It is common to relate the mass of TNT to an overpressure at a given distance by means of an intermediate parameter called the scaled distance. This scaled distance is the actual distance from the explosion divided by the cube root of the mass of TNT involved. Because of the scaling which applies for TNT explosions of different masses, the overpressure is the same for any given scaled distance.

Conceptually this method is capable of wide application as the energy source is not restricted to combustion energy and the method may be used, for example, to model the overpressure effects from a bursting pressure vessel or a rapid phase transition explosion. However, the advantage of this wide applicability also leads to a disadvantage. This is because once an equivalent mass of TNT has been assumed, the remaining parts of the assessment implicitly assume that the characteristics of the explosion and the way it affects its surroundings are independent of the explosion source. Whilst the assumption is satisfactory for explosions involving solid materials, for vapour cloud explosions the shape of the blast wave and the way it interacts with structures are distinctly different. This can have a marked effect in the near field, though in the far field the process of 'shocking up' means that the shape of the blast wave is more nearly independent of explosion source and hence more closely resembles the blast from a TNT source. TNT equivalence modelling continues to be applicable to a wide range of assessments, although caution is needed if it is to be used to predict the near field effects of gaseous explosions.

HIGH EXPLOSIVES

TNT equivalence is most appropriately applied to this class of explosion and the estimation of an equivalent mass of TNT is at its most straightforward. Explosive energy determined, for example, using the lead block test or ballistic mortar can be used to relate an equivalent mass of TNT to the mass of the substance being assessed. Tables 4.2 and 4.3 are included for explosives and reactive/unstable substances, based on such test data.

In predicting explosion hazards from high explosives, it is usual to assume an efficiency factor of unity, even if, during an accidental initiation of high explosives, it may not always release the total explosion energy due to imperfect initiation or confinement.

ENERGETIC MATERIALS

In general, for accidents involving energetic materials, about 0.10–0.15 of the available energy is released in the form of a blast wave, with yields as high as 0.25 being occasionally achieved. It is suggested that this latter value of explosion efficiency should be used.

TABLE 4.2
Energy ratio factors for explosives

Substance	Energy ratio factor (TNT = 1)
Nitroglycerine	1.69
PETN	1.77
Pictric acid	1.06
RDX	1.57
Trinitrobenzene	1.11

TABLE 4.3
Energy ratio factors for energetic materials

Substance	Energy ratio factor (TNT = 1)
2,4 Dinitrotoluene	0.77
Ammonium nitrate	0.56
Cyclohexanone peroxide	0.19
Sodium chlorate	0.15
Nitromethane	1.34

PRESSURE VESSEL EXPLOSIONS

Loss of containment of gas kept under pressure in a vessel can lead to a significant release of energy from the expanding gas. This is true for all gases — inert, toxic or flammable. Release of flammable gas obviously presents a secondary explosion hazard due to combustion which is dealt with in the next section as a 'vapour cloud explosion'.

For the release of stored energy in a pressure vessel, it is difficult to apportion the energy liberated between fracture energy, blast wave energy, energy to project missiles, etc. If the vessel contains a permanent gas (that is one that does not become liquid when compressed) then 40–70% of the available energy may be converted into blast wave energy depending on whether ejection of a major vessel section or fragment occurs. It is suggested that 50% is used as the efficiency factor to cover all cases. Two significantly different cases arise with pressure vessel explosions. The first is that of the so-called permanent gases and the second is gases such as propane and chlorine that may be liquefied under pressure. In the latter case the assessment is complicated by the fact that both gas and liquid are usually present in the vessel and further energy is available from the rapidly vaporizing liquid.

For the permanent gases the energy available from the expanding gases may be calculated from the following formulae:

(a) Brodes equation [1].

$$E = V_1 \, (P_1 - P_a) \, / \, (\gamma - 1)$$

where:

E = energy release;
V_1 = original volume of gas;
P_a = pressure of surrounding atmosphere;
P_1 = original pressure of gas;
γ = ratio of specific heats of gas (assumed constant).

(b) Brinkley's equation [2].

$$E = \frac{(P_1 V_1)}{\gamma - 1} \, [1 - \, (P_a \, / \, P_1)^{(\gamma - 1)/\gamma} \,]$$

In general the second formula is to be preferred.

Once the energy available has been calculated, it is converted to an equivalent mass of TNT by dividing it by the energy content of TNT, which is 4.6 MJ $kg^{-1.}$

The shape of the blast wave can be significantly different from an ideal blast wave from TNT and the results should be used with caution. An alternative method is described in Section 4.7 (see page 40).

When a vessel contains a liquefied compressed gas, some of the liquid might vaporize (flash off) following rupture of the vessel. A suitable equation to calculate the energy available, to be converted into a TNT equivalence, is as follows:

$$E = \frac{P_1 \, (V_1 + V_2)}{\gamma - 1} \, (1 \, (P_a \, / \, P_1)^{(\gamma - 1)/\gamma})$$

where V_1 is the vessel volume and P_1 the pressure in the vessel and

$$V_2 = \frac{22.4\,F\,M}{MW}\,(P_1\;/\,P_a)$$

where:
MW = molecular weight;
M = mass of liquid in the vessel;
F = flash fraction (depending on the temperature).

VAPOUR CLOUD EXPLOSIONS

Since the 1970s, when there were several disastrous vapour cloud explosions (VCEs) of this type, a considerable amount of attention and research effort has been directed at this subject. Immediately after this period the UK Advisory Committee on Major Hazards (ACMH) considered the implications of major hazard sites for the community as a whole, and the control and mitigation regimes that might be applied to them. Part of the Committee's deliberations covered how the effects of unconfined vapour cloud explosions could be assessed.

The Second Report of the Advisory Committee [3] recommended the use of the TNT equivalence method, using certain assumptions, for the assessment of VCEs.

They noted that many factors influence the magnitude and therefore the consequences of a VCE. As the effects of each of these factors is not known with any certainty, an overall judgement is made as to TNT equivalence factor. For hydrocarbons generally a study of observed damage from accidental explosions suggests that the combustion energy of the cloud be multiplied by an efficiency factor of 0.03, or 0.042 if allowance is made for the radius of the flammable cloud which allows distances to be measured from the cloud centre. Since the energy of combustion is about ten times the energy of detonation of TNT, then 1 tonne of hydrocarbon vapour in a flammable cloud is equivalent to 0.3, or 0.42 tonne of TNT (that is, an energy ratio of 10). Some other values are given in Table 4.4. on page 30.

The combustion energy for flammable gases and vapours is available in many references, for example Reference 4.

For other substances it is suggested that the TNT equivalence should be calculated by multiplying a calculated energy ratio factor by one of the efficiency factors given in Table 4.4 according to their perceived reactivity. For hydrocarbons generally an efficiency of 0.04 should be used, for above average reactivity 0.06 and for very reactive 0.1 should be used.

When the fuel is being generated by the release of a flashing liquid, it is a common assumption to calculate the mass of gas produced by the flashing process and then double it to allow for the liquid entrained as aerosol by this energetic physical process.

A method based on TNT equivalence has the great advantage of using the vast amount of data available relating to war and accidental damage effects, although the maximum overpressure of a VCE is less than the maximum overpressure from TNT. It has been recommended that a maximum overpressure for a non-detonating VCE can be assumed to be 1 bar.

The disadvantage of the method lies in the arbitrariness of the efficiency factor.

More recent research concerned with hydrocarbons which are predominantly methane (Section 4.3) has identified that the main source of blast energy comes from regions where there are repeated obstacles. This has resulted in TNT-equivalent methods which are based on the flammable volume within obstructed regions, but with a higher efficiency factor.

RAPID PHASE TRANSITIONS

The TNT equivalent mass for a rapid phase transition (RPT) is proportional to the amount of energy in the superheat of the fluid at the time of the explosion. Only when a liquid becomes superheated (that is, raised to a temperature above its boiling point at atmospheric pressure) can it undergo an RPT. This energy is related to the mass of superheated fluid, the degree of superheat (that is, the

TABLE 4.4
Factors for gases and vapours

Material	Energy ratio factor	Efficiency factor	Equivalence factor
Hydrocarbons	10	0.04	0.4
Ethylene oxide	6	0.1	0.6
Vinyl chloride monomer	4.2	0.04	1.16
Propylene oxide	6.9	0.06	0.4

temperature excess above the boiling point) and thermodynamic properties of the fluid (for example, latent heat of evaporation, heat capacity).

The mass of liquid that can become super-heated is very much dependent upon the way in which the two fluids interact with one another. There is also an upper bound to the amount of superheating of the colder fluid determined by the superheat limit temperature. Superheat limit temperatures vary considerably between different fluids.

Because of these many factors, for which a detailed calculation procedure does not presently exist, it is recommended that specialist advice is sought when estimating blast from RPTs.

As an isolated example, although it is emphasized that the result can not be generalized to other cryogens, a TNT-equivalence of 3.5 kg has been measured[5] in experiments involving a release of 24 m^3 (24×450 kg/m^3 = 10,800 kg) of liquified natural gas on to water. The interaction between the two fluids was forced, at a discharge rate of 1086 m^3/hr.

4.3 BRITISH GAS VCE METHOD

The British Gas method is intended for non-detonating clouds of natural gas (methane with up to 5% ethane).

The experimental results[6] demonstrate the importance of flame acceleration in congested regions, stimulated by repeated obstructions, in determining the overpressures generated by a VCE. For potential releases of natural gas, the results described in Reference 6 have been used to develop an improved method for the prediction of pressures. This is not based on the entire mass of flammable material released, but on the mass of material which can be contained in stoichiometric proportions in any severely congested region of plant. The experiments reported have demonstrated that, under particular conditions, higher overpressures can be generated than were suggested by the ACMH[3]. The method assumes, pessimistically, that within the exploding part of the vapour cloud an overpressure of 4 bar will be produced — the maximum measured in any of the experiments. The efficiency of energy release used in the TNT equivalent calculation is also increased, to 0.2. Using an energy ratio of 10 for hydrocarbon-air explosions, the TNT equivalent mass is therefore given by:

$$M_{\mathrm{TNT}} = 2M_{\mathrm{congested}}$$

where $M_{congested}$ is the mass of natural gas which can be contained in stoichiometric proportion mixed with air, in any severely congested region of plant which a vapour cloud envelops. For any hydrocarbon-air vapour cloud, a given volume will contain approximately the same mass of flammable vapour (0.08 kg/m^3). Thus the above equation can be simplified to:

$$M_{TNT} = 16\,V \times 10^{-2}\,\text{kg}$$

where V is the smaller of either the total volume of the congested region or the volume of the entire vapour cloud, and is measured in m^3.

To satisfy the criterion that the overpressure within the exploding part of a vapour cloud is 4 bar, as the equivalent TNT mass is a point source, it has to be located at a position within this region such that the magnitude of the decaying pressure wave is 4 bar at the cloud edge. The scaled distance at which an overpressure level of 4 bar occurs is 1.6. Thus, the overpressure at a distance r metres from the edge of a congested region can be determined from Figure 3.2 using a 'scaled distance' given by:

$$sd = 1.6 + r\ /\ M_{TNT}^{\ 1/3}$$

Where the consequences of large potential releases of natural gas have to be considered on sites where there is very little congestion, the revised method predicts significantly lower far-field overpressures. In many circumstances (such as those where any cloud formed would not engulf regions of repeated obstacles) the possibility of VCEs with significant overpressures involving natural gas can be discounted.

For releases of gases considered more reactive than natural gas, the experiments have demonstrated that where part of the cloud can engulf a severely congested region, it may not be appropriate to apply a methodology based on an efficiency factor of 0.2 and an overpressure level within the congested region of 4 bar. Under specific circumstances the possibility of flame acceleration leading to detonation of part of any cloud might have to be considered and a different method used.

4.4 TNO MULTI-ENERGY METHOD

Van den Berg [7,8], working in the Prins Maurits Laboratory of TNO, Rijswijk, The Netherlands, also considered that the major component of blast originated in congested regions. In his method the relationship between overpressure and scaled distance is based on computed calculations of the pressure decay following a gaseous explosion [8]. In Reference 7 the method is named the TNO multi-energy method.

This method assumes a quiescent, hemispherical cloud of stoichiometric concentration, equal in volume to the region of cloud under consideration. Blast parameters are calculated using a 1–D constant flame speed explosion model based on numerical integration of the compressible flow equations. To simplify the use of the method, the results from the explosion model are presented as plots of dimensionless blast parameters against combustion energy–scaled distance for different initial blast strengths (see Figures 4.1, 4.2 and 4.3). It is then a case of estimating the initial blast strength, as indicated by a number ranging from 1 for insignificant strength to 10 for a gaseous detonation, and reading the overpressure-distance data from Figure 4.1, and impulse and positive phase duration from Figures 4.2 and 4.3 (see pages 34 and 35). The method assumes that strong blast is generated only by that part of

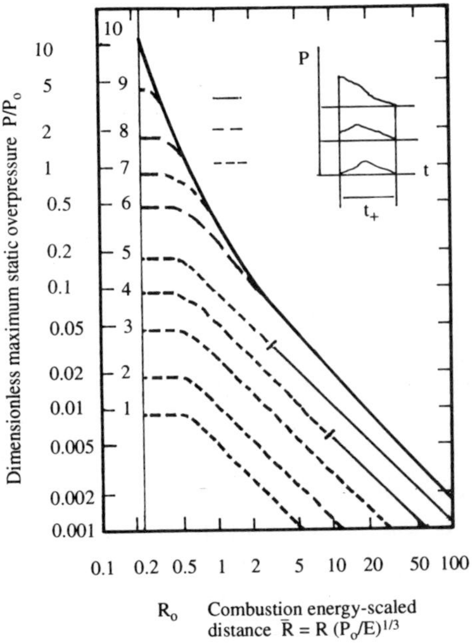

Figure 4.1 TNO multi-energy method: overpressure. (van den Berg [7].) (Reproduced by permission of Elsevier, Amsterdam, The Netherlands.)

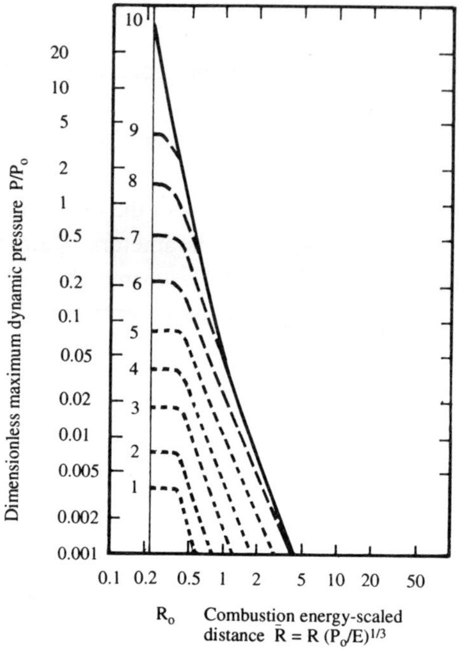

Figure 4.2 TNO multi-energy method: impulse. (van den Berg [7].) (Reproduced by permission of Elsevier, Amsterdam, The Netherlands.)

the cloud that is subjected to confinement and congestion, while the remaining parts of the cloud just burn out with no significant contribution to the blast. Therefore, to predict the blast wave parameters, the unconfined and confined/congested parts of the cloud are modelled separately, using a high value for the initial blast strength for the confined/congested part and a low value for the unconfined part, and combine to give the overall result. If there are two or more areas of confinement/congestion then these should be modelled separately. Since flame propagation in the unconfined parts of the cloud is relatively slow it is unlikely that two strong sources of blast will be initiated at the same time, and the blast waves from the two sources can be treated independently of each other.

Overpressures at any position are estimated from explosion in each congested region and from explosion in the uncongested cloud volume. The

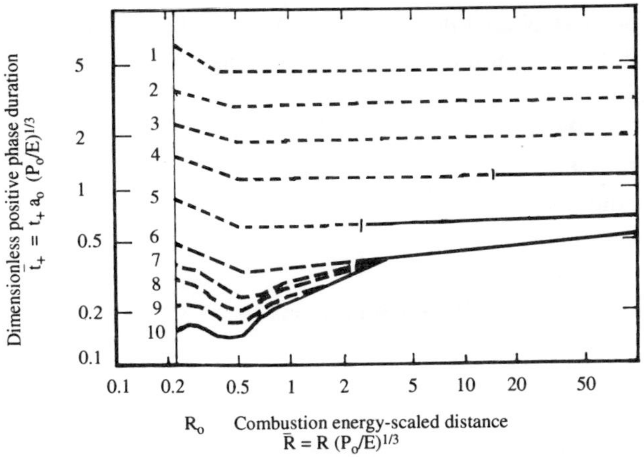

Figure 4.3 TNO multi-energy method: positive phase duration. (van den Berg[7].) Reproduced by permission of Elsevier, Amsterdam, The Netherlands.)

final estimate of overpressure is the maximum of the contributions from each congested region and from the uncongested part of the cloud.

Assessment using the TNO multi-energy method typically proceeds as follows:

- From the geometry of the hazard area, calculate the total volume, V_1, of each obstructed/congested region. Calculate V_2, the largest volume of stoichiometric cloud the stored inventory of flammable material is capable of producing. If V_2 is less than V_1, take V_1 as V_2. Calculate $V_3 = V_2 - V_1$.
- Calculate the energy, E, stored in stoichiometric clouds of volumes V_1 and V_3 if non-zero.
- To find the overpressure at a given distance, R, for V_1 and V_3, calculate the 'combustion energy-scaled distance', by dividing R by the cube root of E, and then multiplying by the cube root of ambient pressure (in pascals).
- Read off the resulting overpressure as a function of 'combustion energy-scaled distance' from one of the calculated curves in Figure 4.1. A typical recommendation is to use curve 7 (corresponding to 1 bar) for the confined part of the cloud, curve 2 (corresponding to 20 millibar) for the unconfined part of the cloud, and add up the resultant pressures. TNO also provides the dynamic pressure (Figure 4.2) and the duration of the positive

phase (Figure 4.3). The impulse is one half of the product of the peak overpressure and the positive phase duration for a triangular shaped wave.

Figures 4.1, 4.2 and 4.3 use the following symbols:

P_s = overpressure, Pa;
P_0 = ambient pressure, Pa;
P_d = dynamic pressure, Pa;
t_+ = positive phase duration, s;
c_0 = ambient velocity of sound, m/s;
E = combustion energy involved;
R_0 = energy–scaled radius of the obstructed region.

4.5 SHELL METHOD

Cates, working at Shell's Thornton Research Laboratory, has proposed a further multi-energy method [9]. Cates considered that overpressure in the extreme far field was underestimated by earlier methods. The basic principle is the same in that the major component of blast is generated only in obstructed regions of plant, but it differs in that the overpressure is calculated from a simple inverse decay law. The method is linked to a decision tree (Figure 4.4), from Reference 9, to guide the user through the decisions needed to estimate a source overpressure.

The decision tree takes some of the arbitrariness out of guessing source overpressures, although any estimates are uncertain and subject to future revision. Some judgement from the user is required in using the decision tree.

The term 'bang box' mentioned in the decision tree refers to an ignition within a building or box with venting of a high velocity flame through windows or through other orifices in the walls or roof. Such burning jets form a very strong ignition for a vapour cloud and can lead directly to rapid flame propagation and high overpressure. With vigorous 'bang box' ignition, detonation may need to be considered.

The cloud is divided into obstructed and unconfined parts. A volume, V_1, is derived as an estimate for the volume of burnt and compressed gas generated by the explosion. Detailed consideration of the geometry can often reduce the volume which should be taken into account, but by default twice the total volume of gas within 2 m of any obstacle should be calculated. It is assumed in subsequent calculations that the volume is hemispherical, and its radius, R_0, is calculated.

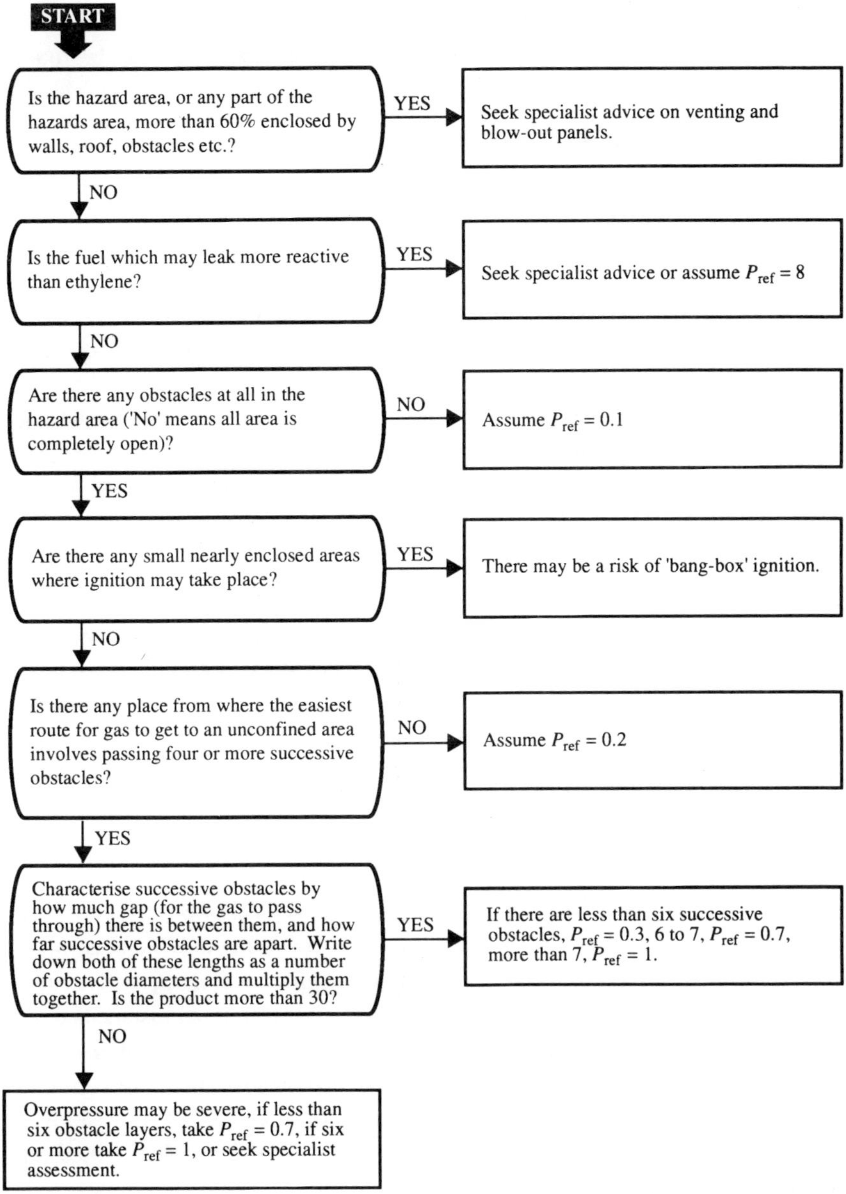

Figure 4.4 Decision tree.

At a distance R from the edge of the hazard area (boundary of the hemisphere, with a radius of R_0), the peak overpressure, P_{max}, can be found from:

$$\text{Overpressure at } R = R_0 (P_{max} / P_{ref} - 1)$$

where:

$$R_o = (3V_1/2\pi)^{1/3}$$

The method to calculate overpressures proceeds as follows:

- From the geometry of the hazard area, calculate the total volume of explosion products, V_1, within 2 m of any group of obstacles, and then double that estimate. Calculate V_2, the largest volume of stoichiometric cloud which the stored inventory of flammable material is capable of producing. If V_2 is less than V_1, take V_1 as V_2. Calculate a similar volume representing the effects of the unconfined cloud, $V_3 = V_2 - V_1$. Separately calculate the radii of the hemispheres containing V_1 and V_3, being R_1 and R_3.
- For each volume, V_1 and V_3, estimate a reference pressure, P_{ref}, using the decision tree in Figure 4.4. Multiply P_{ref} by 2.5 if the fuel gas is ethylene. If the gas is more reactive than ethylene, seek specialist advice or assume $P_{ref} = 8$.
- At a distance R from the edge of the hazard (R_1 and R_3) estimate the overpressure from each part of the cloud, using:

$$\text{Overpressure at } R = R_0 (P_{max} / P_{ref} - 1)$$

4.6 FLAT (PANCAKE SHAPE) AND SLENDER CLOUDS

Most of the methods described in Sections 4.2–4.5 assume that the blast wave evolves symmetrically from a single point on the ground. Consequently, equal overpressures are predicted to occur at all points lying at equal distances from the 'explosion centre'. The centre corresponds to the position of the high explosive charge for the standard TNT equivalence method and to the centre of the presumed hemispherical vapour cloud for the TNO Prins Maurits Laboratory and Shell methods. The British Gas TNT–equivalence method, however, differs by defining the explosion centre to have different positions which depend on the shape of the vapour cloud boundary. The predicted overpressure, therefore, ceases to be symmetrical about the centre of the cloud when the base of the cloud is not circular. An alternative treatment of non-hemispherical clouds, suggested by Cates[9], is to break down the total volume into a number of separate hemispheres and then to add the overpressures for each separate hemisphere together. These latter methods have been developed to address, albeit in a way which yields conservative predictions, the more difficult question of predicting the blast wave from vapour clouds of general shape.

For detonation, or very high flame speeds, it has been shown by numerical studies[10], that the far-field blast from a pancake cloud is closely approximated by the TNT-equivalence method. This assumes a TNT efficiency factor appropriate to detonation, which for hydrocarbon-air mixtures is approximately 0.4. In the near-field, however, which can be defined as the region adjacent to the cloud boundary, the overpressure and impulse can be substantially higher than that given by the TNT method. The same high near-field blast characteristics have also been observed experimentally for limiting flame speeds[11], sometimes referred to as quasi-detonation, propagating through slender congested regions following a high energy ignition.

For flames with a sufficiently low speed, a very different conclusion has been reached[12,13]. The dominant far-field blast is equal to that given by a centrally ignited hemispherical cloud located such that the hemisphere is tangent to the cloud boundary, with the equivalent hemisphere radius equal to the cloud height.

Once the flame ceases to be hemispherical, either as a result of its extinction at the boundary of the flammable mixture (at the upper limit of the cloud or at its lateral limit) or at the boundary of a congested region, not only does the flame decelerate, but also the blast overpressure generated subsequently is significantly lower than that of the hemispherical flame. Thus, that phase of the explosion in which the flame has a hemispherical shape gives rise to the dominant blast.

The TNT methods give a conservative estimate of the blast. However, in certain cases, namely those in which the maximum flame speed is known to be below a certain threshold level, the explosion efficiency is very much lower for high aspect ratio clouds. In these cases a hemispherical flame model such as in the TNO Prins Maurits Laboratory or Shell methods can be employed to predict far field overpressure but where the cloud radius is now assigned a smaller value based on the reasoning above. It is recommended that any decision to use the 'reduced radius' hemispherical cloud approach for non-hemispherical clouds should be substantiated by a more detailed analysis than that provided in this text and specialist advice should be sought.

4.7 BURSTING PRESSURE VESSELS

COMPRESSED GASES

Systematic calculations of the overpressure from a bursting pressure vessel were made by Strehlow and Ricker[15], later reported by Baker *et al*[16]. The calculations included the effects of variations in vessel internal pressure and temperature, and in specific heat and the ratio of specific heats. The results of the calculation were summarized as plots of scaled overpressure versus scaled radius, Figure 4.5. To use the graphs to predict the overpressure at any radius requires that the appropriate line or interpolation be identified from estimation of an initial dimensionless pressure and an equivalent dimensionless sphere radius. The equations to be used are:

$$\frac{P_1}{P_0} = \frac{P_{so}}{P_0}\left(1 - \frac{(\gamma_1 - 1)(a_0/a_1)\,\{(P_{so}/P_0) - 1\}}{(2\gamma_0\,\{2\gamma_0 + (\gamma_0 + 1_0)(P_{so}/P_0 - 1)\})^{1/2}}\right)^{\frac{-2\gamma_1}{\gamma_1 - 1}}$$

where solution for P_{so} / P_0 is found by iteration, and the initial dimensionless overpressure, $\overline{P}_{so}$ is $P_{so} / P_0 - 1$

P_1 = vessel pressure;
P_0 = atmospheric pressure;
P_{so} = initial shock pressure for the blast wave;
$\overline{P}_{so}$ = initial dimensionless overpressure;
a_1 = speed of sound in compressed gas;
a_0 = speed of sound in atmosphere;
γ_1 = ratio of specific heats in compressed gas;
γ_0 = ratio of specific heats in atmosphere.

The energy contained in the vessel, volume V_1, is:

$$E = V_1\,(P_1 - P_0)\,/\,(\gamma_1 - 1)$$

and the equivalent dimensionless sphere radius, $\overline{R}_1$:

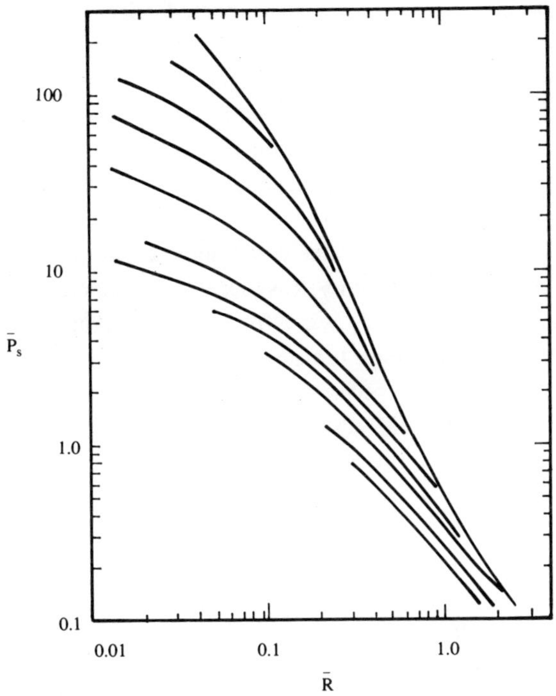

Figure 4.5 Overpressure from a bursting pressure vessel containing compressed gas. (Baker et al [15].) (Reproduced by permission of Elsevier, Amsterdam, The Netherlands.)

$$\overline{R}_1 = [3\,(\gamma_1 - 10 / \;\{\,4\pi\,(P_1 / P_0 - 1)\,\}\,]^{1/3}$$

To use this method, simply locate $\overline{P}_{so}$ and $\overline{R}_1$ on the appropriate curve of Figure 4.4. Overpressures at a greater radius are found from following the curve to higher values of scaled equivalent radius.

The actual overpressure P is found from:

$$P_2 = \overline{P} / P_o$$

and the actual radius from:

$$R = \overline{R}\;(E / P_0)^{1/3}$$

LIQUEFIED COMPRESSED VAPOURS

When the liquid content of a pressurized vessel is overheated relative to its boiling point at atmospheric pressure, then rupture of the vessel can result in a blast wave. The blast is generated by the rapid boiling and expansion in the volume of the liquid contents which occurs upon depressurization.

If the vessel is not completely full of liquid, the expansion of the vapour is an additional potential source of blast which should be evaluated using the above method for the gas-phase expansion.

A methodology for calculating the blast due to the liquid expansion has been proposed by the TNO Prins Maurits Laboratory, Rijswijk, The Netherlands [16]. This is based upon measurements of the blast wave from vessels containing liquid propylene [17] at different overheat temperatures. The overheat temperatures (*dt,* Kelvin) used in these experiments (see Figure 4.6) are for propylene which has to be modified (*DT*, Kelvin) for other liquids.

Firstly the 'modified overheat temperature' is calculated from:

$$DT = dt \times 0.007 \ / \ SLR$$

where *SLR* (per degree Kelvin) is the ratio of heat capacity to latent heat of evaporation for the liquid.

Secondly the scaled blast radius (R) is calculated from the distance from the centre of the vessel (r, metre) and the initial mass of liquid (M_{liquid}, kg), taking into account the effects of blast wave reflection from the ground:

$$R = r \ / \ (2M_{\text{liquid}})^{1/3}$$

The incident peak overpressure is then read off the profile in Figure 4.6 for the relevant overheat temperature. The methodology recommends that the positive impulse be calculated assuming a triangular profile for the pulse with a duration (t, secs) given by:

$$t = 2.94 \times 10^{-3} \times r$$

Note that close to the vessel the blast wave will comprise a two-phase mixture (see Figure 4.6), and the above methods for calculating the impulse may no longer be valid.

4.8 MISSILES

The energies of primary missiles can be estimated using a number of procedures reported in the literature [15,18,19,20]. A table of energy levels for impact design can

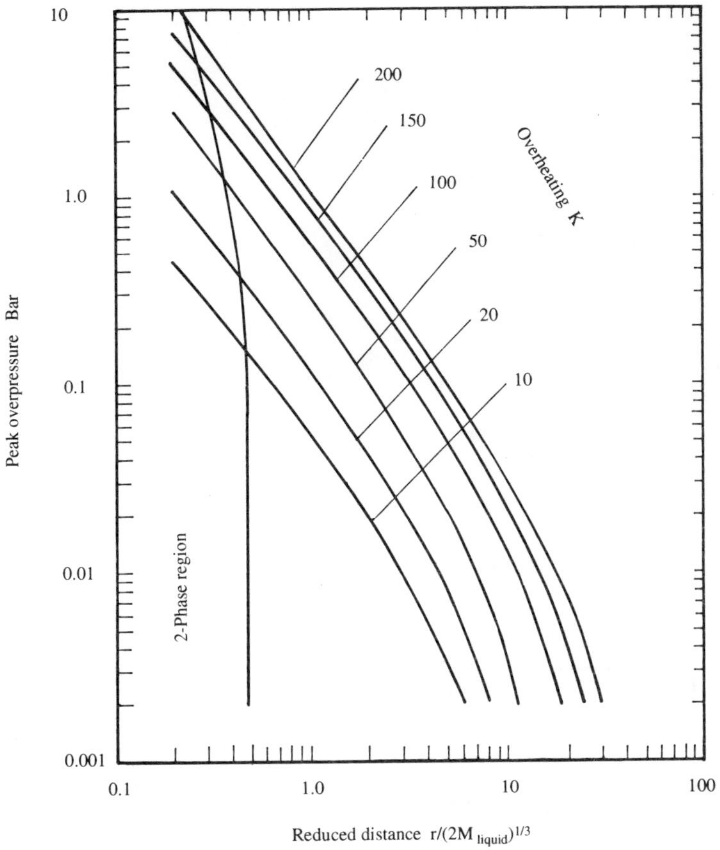

Figure 4.6 Pressure decay from the flashing liquid shock. (Maurer [17].) (Reproduced by permission of the DECHEMA c.V., Frankfurt am Main/Germany. Proceedings, "2nd International Symposium on Loss Prevention and Safety Promotion in the Process Industries", 1978 (ISBN 3 921567 11 4).)

be found in Reference 20. They range from several kilojoules up to tens of thousands of kilojoules. The table shows that the maximum possible energy of primary projectiles is likely to considerably exceed the ultimate strength of any practical and viable impact protection system [21].

A procedure to evaluate the velocities of missiles entrained by blast waves (secondary missiles) is given by Baker *et al* [15]. The velocity of a missile is given by:

$$V = \overline{V} P_a \; A \; (KH = X) \; / \; (M a_a)$$

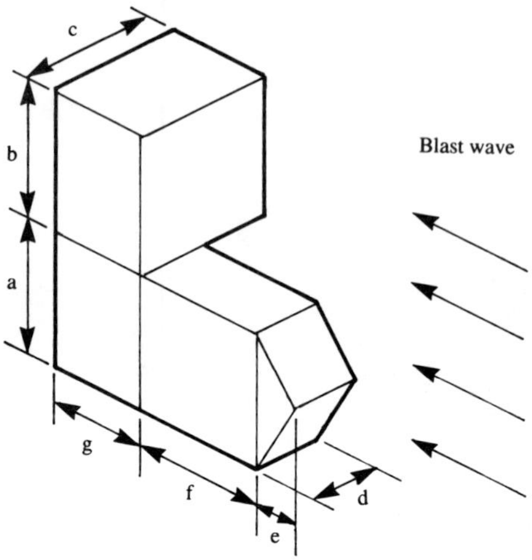

Isometric view of a potential missile

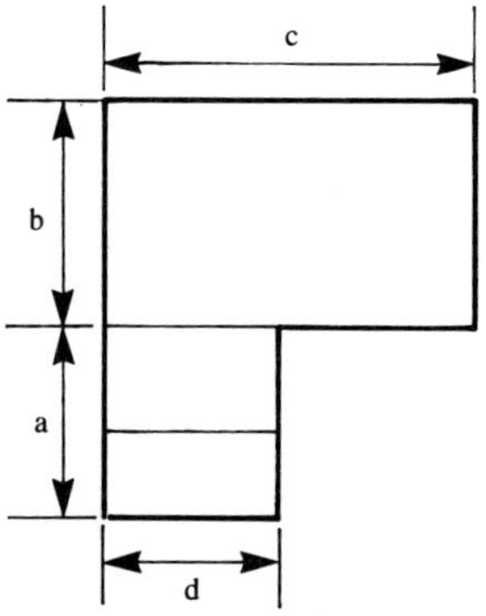

Front view of a missile

Figure 4.7 Definition of missile dimensions. A = bc + ad, H = d, X = e + f.

where:

M = mass of missile;

A = projected area of missile;

K = a constant, takes a value of 4 if the potential missile is on the ground and 2 if it is in the air;

P_s = blast wave overpressure;

I = blast wave impulse;
P_a = ambient pressure;
a_a = velocity of sound in air;
H = the minimum transverse dimension of the presented area of the potential missile;
X = the distance from the front of the potential missile to its largest cross-sectional area;
$\overline{V}$ = non-dimensional velocity (see Figure 4.7).

$\overline{P}$ = P_s / P_a

$\overline{I}$ = $C_d I a_a / [P_s (KH + X)]$ (see Figure 3.1, see page 17)

(values of $\overline{P}$ and $\overline{I}$ are used to locate $\overline{V}$ from Figure 4.8)

C_d = drag coefficient (see Figure 4.9, page 46)

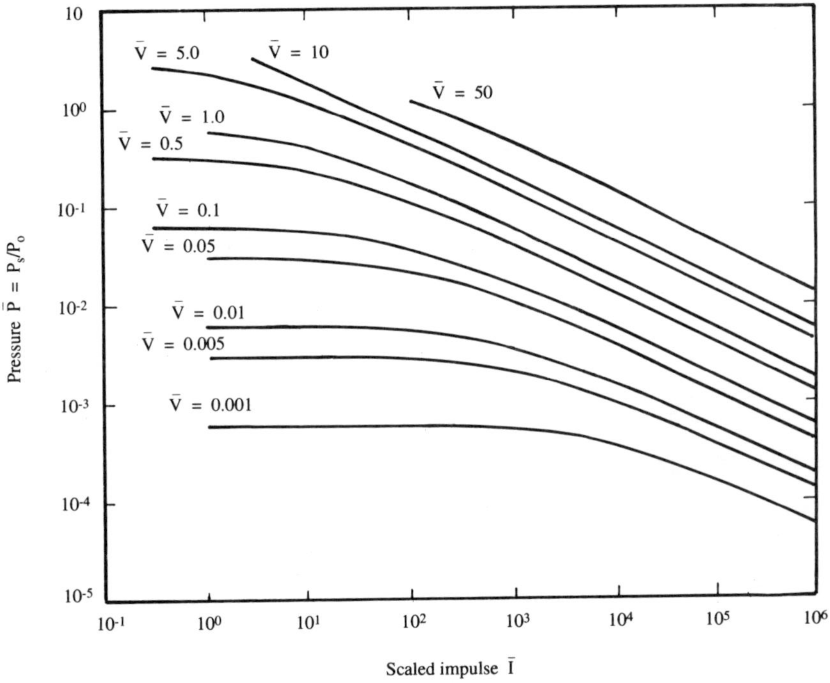

Figure 4.8 Missile velocities. (Baker et al [15].) (Reproduced by permission of Elsevier, Amsterdam, The Netherlands.)

Shape	Sketch	C_D
Right circular cylinder (long rod), side-on	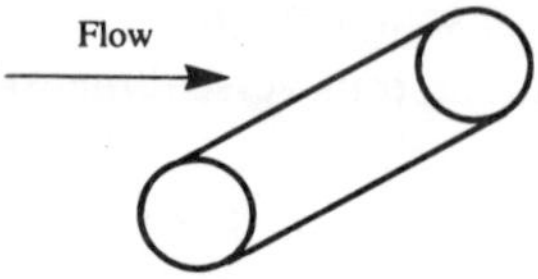	1.20
Sphere	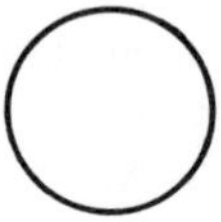	0.47
Rod, end-on	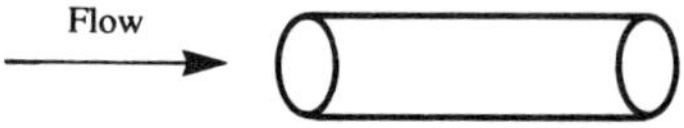	0.82
Disc, face-on	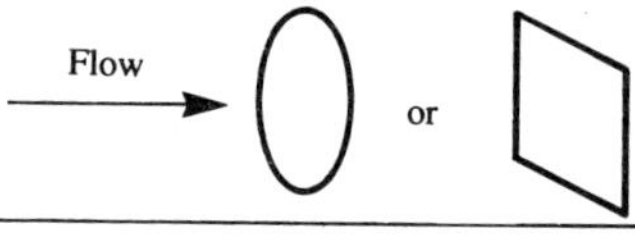	1.17
Cube, face-on	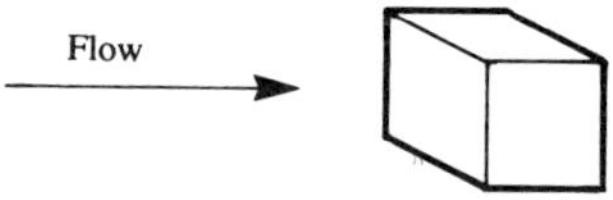	1.05
Cube, edge-on	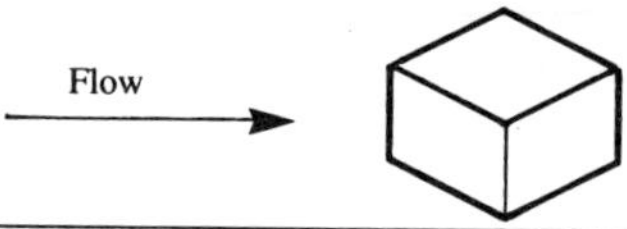	0.80
Long rectangular member, face-on	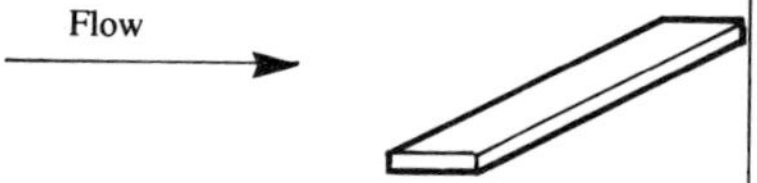	2.05
Long rectangular member, edge-on	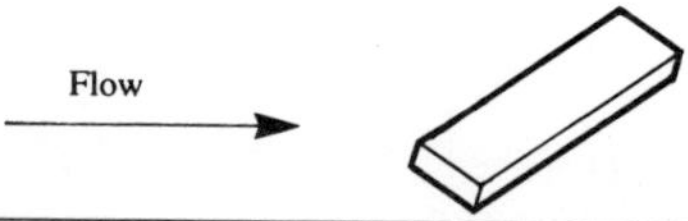	1.55
Narrow strip face-on	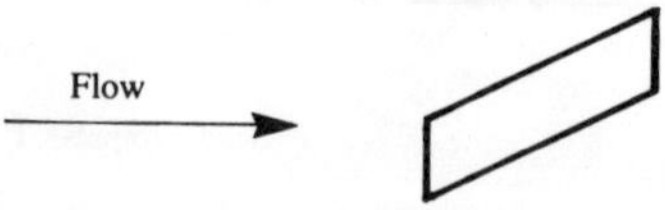	1.98

Figure 4.9 Drag coefficients. (Baker *et al* [15].) (Reproduced with permission from Elsevier, Amsterdam, The Netherlands.)

Tam and Simmonds[22] proposed a model to calculate the drag force, F_d, on an object in the path of the blast wind, which is given by:

$$F_d = C_d A [0.5 \rho (V_{gas} - V_{mis})^2]$$

where:

C_d = drag coefficient;
A = projected area of missile;
ρ = density of gas;
V_{gas} = velocity of gas;
V_{mis} = velocity of missile;

The following equation was then derived for the maximum velocity of an entrained missile:

$$V_{mis,max} = V_{gas,max} [1 - 2^{0.5} \arctan (\Phi)]$$

where:

a^2 = $0.5 C_d A \rho V_{gas,max} t_{max}/M_{mis}$
Φ = $\tanh (a/2)^{0.5})/a$;
t_{max} = width of the velocity time curve;
M_{mis} = mass of the missile.

REFERENCES IN CHAPTER 4

1. Brode, H.L., 1959, *Physics of Fluids*, 2: 217.
2. Brinkley, S.R., 1969, Determination of explosive yields, *Loss Prevention 3* (AIChE), 78–82.
3. Health and Safety Commission, 1979, *Advisory Committee on Major Hazards,* Second Report, (HMSO, UK).
4. Spiers, H.M., Technical data on fuel, *The British National Committee World Power Conference, London, UK.*
5. McRae,T.G., Goldwire Jr., M.C. and Koopman, R.P., 1984, Analysis of large scale LNG/water RPT explosions, *Paper WA/HT–76* (ASME).
6. Harris, R.J. and Wickens, M.J., 1989, Understanding of vapour cloud explosions—an experimental study, *55th Autumn Meeting, Communication 1408* (Institute of Gas Engineers, UK).
7. van den Berg, A.C., 1985, The multi-energy method. A framework for vapour cloud explosion blast prediction, *J Haz Mat*, 12,: 1.

flux-corrected transport algorithm (Prins Maurits Laboratory TNO, Report PML)
9. Cates, A.T., 1991, *Fuel Gas Explosions Guidelines: Fire and Explosion Hazards, Energy Utilisation* (Institute of Energy, IOP Publishing, UK).
10. Raju, M.S. and Strehlow, R.A., 1984, Numerical investigations of non-ideal explosions, *J Haz Mat*, 9: 265–290.
11. Harris, R.J. and Wickens, M.J., 1989, Understanding of vapour cloud explosions — an experimental study, *55th Autumn Meeting, Communication 1408* (Institute of Gas Engineers, UK).
12. Catlin, C.A, 1985, An acoustic model for predicting the overpressures caused by the deflagration of a ground lying vapour cloud, *IChemE Symposium Series No 93, Assessment and Control of Major Hazards* (IChemE, UK).
13. Rosenblatt, M. and Hassig, P.L., 1986, Numerical simulation of the combustion of an unconfined vapor cloud at a high constant burning velocity, *Comb Sci and Tech*, 45: 245–259.
14. Strehlow, A. and Ricker, R.E., 1976, The blast wave from a bursting sphere, *Loss Prevention* 10: 115–121 (AIChE).
15 Baker, E., Cox, P.A., Westine, P.S., Kulesz, J.J. and Strehlow, A., 1983, *Explosion Hazards and Evaluation* (Elsevier, Amsterdam, The Netherlands).
16. *Methods for the Calculation of the Physical Effects of the Escape of Dangerous Material, Report of the Committee for the Prevention of Disasters*, Edition 1, 1979 (Ministry of Social Affairs, Balen van Andelplein 2–2273 KH Voorburg, The Netherlands).
17. Maurer, B., Hess, K., Giesbrecht, H. and Leuchel, W., 1977, Modelling of vapour cloud dispersion and deflagration after bursting of tanks filled with liquefied gas, *2nd Intl Symposium on Loss Prevention and Safety Promotion in the Process Industries*.
18. Baum, M.R., 1987, Disruptive failure of pressure vessels: preliminary design guidelines for fragment velocity and the extent of the hazard zone, *Recent Advances in Structural Dynamics*, 124: 175–181 (ASME Publication PVP).
19. Baker, W.E., Cox, P.A., Westine, P.S., Kulesz, J.J. and Strehlow, R.A., 1983, *Explosion Hazards and Evaluation* (Elsevier, Amsterdam, The Netherlands).
20. Design of structures to resist nuclear weapons effects, *ASCE Manual* 42.
21. *Interim Guidance Notes for the Design and Protection of Topside Structures Against Explosion and Fire*, 1991.
22. Tam, V.H.Y. and Simmonds, S.A., 1990, Modelling energy from gas explosions offshore, *International Conference on the Management and Engineering of Fire Safety and Loss Prevention Onshore and Offshore*.

5. ESTIMATION OF INJURIES AND DAMAGE

5.1 INTRODUCTION

For an accurate estimate of the injuries and damage caused by explosions, the following three stages are involved:

- predicting the time history of the blast wave produced in the air, for both the overpressure and the dynamic pressure;
- predicting the time history of the load acting on the person or structure as a result of interaction with a blast wave or missile;
- predicting the response of the structure or person to this loading.

However, a more pragmatic approach for predicting the likely consequences of an explosion accident with sufficient accuracy for a preliminary assessment is to determine the overpressure and to consult tabulated data to estimate the likely damage.

The reader is referred to Chapter 3 for a review of blast wave terminology and a general description of how blast waves interact with objects and cause injuries and damage. Chapter 4 discusses the methods that are available for predicting the overpressure and dynamic pressure time histories. In the rest of this chapter the methods available for predicting the loading and response of objects and missile damage are presented.

5.2 RESPONSE OF STRUCTURES TO BLAST

In estimating the loading on a structure it should be born in mind that the blast wave impinging on the structure can be significantly different from the blast wave propagating through undisturbed air — that is, the free air parameters. This results from reflection of the blast wave from the ground, other structures or even adjacent parts of the structure. The prediction methods given in Chapter 4 will predict the free air blast parameters.

Methods for calculating the blast wave properties after interaction with simple structures and the resulting loading can be found in Glasstone and Dolan [1] and Baker [2]. These methods should strictly be applied only to shock waves and, in the case of gas and vapour explosions, well outside the region of combustion. In the case of pressure waves rather than shock waves, similar methods can be applied with a reasonable degree of accuracy, provided information is available on the densities and air velocities within the blast wave. For geometries other than very simple ones these methods become increasingly difficult to apply.

A number of computer codes are available for calculating blast wave interactions with structures, but their use is beyond the scope of this monograph. The reader is referred to Reference 3 for further information on these models and the calculation of blast loading.

It is usual in blast loading calculations to assume a constant drag coefficient and to use coefficient values obtained under steady flow conditions. However, the value of the drag coefficient depends on the flow velocity and the air velocities produced by blast waves are far from steady. A discussion on drag coefficients and effects of transient velocities can be found in Reference 3. For predicting off-site hazards, which is the main area of interest, the assumption of a constant drag coefficient is reasonable. At some distance from the explosion centre the 'wavelength' of the blast wave is large compared to the structure, so allowing any transient effects due to a changing velocity to be neglected.

Information on drag coefficients to use under this assumption can be found in Hoerner[4] and Uppard[5].

A more detailed discussion on the response of structures to a blast wave, and the relative importance of overpressure and impulsive loading, can be found in Reference 3.

As noted in Section 5.1, an alternative and more pragmatic approach to calculating loadings and responses for predicted blast wave parameters is to use the experience gained from analysis of past incidents or the results of experimental studies to estimate likely damage patterns. For high explosives there is a considerable body of data obtained from military tests and analysis of war damage. From this data tables such as Table 5.1, for TNT, can be produced[6]. The differences in overpressures for different masses of explosive effectively illustrate the importance of impulse (or duration) in assessing blast damage, the duration and impulse increasing with mass of explosive for a given peak overpressure. In many cases overpressure versus damage data is published without any indication of the mass of explosive and thus the duration and impulse values.

Caution is required in applying data, such as given in Table 5.1, in assessing explosion damage from other types of explosion — for example, from

TABLE 5.1
Effect of explosives quantity on failure criteria

Structural element	Failure mode	Approximate peak side-on overpressure (kPa) at which failure occurs for 3 sizes of TNT charge		
		1 tonne	10 tonne	100 tonne
Window pane	5% broken	1.0	0.7	0.7
	50% broken	2.5	1.7	1.5
	90% broken	6.2	4.1	3.7
House	Tiles displaced	4.4	2.9	2.6
	Doors and window frames broken in	9.0	5.9	5.3
	Minor structural damage	6.2	4.1	3.7
	Partial or total collapse of roof	27.6	16.5	15.9
	50 to 70% external brickwork destroyed or rendered unsafe	79.3	35.9	34.5
	Almost complete demolition	183	79.3	75.8
Telegraph pole	Snapped	358	179	165
Large trees	Destroyed	393	179	165
Rail wagons	Limit of derailment	26.5	11.5	11.0

vapour cloud explosions. Due to the differences in the blast waves produced by the two types of explosion this could lead to some very misleading damage assessments, particularly near the explosion centre where the overpressure from a vapour cloud explosion is much less than that from a TNT explosion. At some distance from the explosion the overpressures in blast waves from different types of explosion do tend towards a similar shape, although the impulse might differ. Therefore, Table 5.1 and other data obtained with high explosives can be used for rough estimates of damage levels in the far field for all types of explosion. Table 5.1 shows the effect of the greater impulse in larger explosions of TNT. A vapour cloud explosion has an impulse larger than a TNT explosion (for the same overpressure) so it is recommended that the data for the larger TNT explosions be used for vapour clouds.

There are a few sources of blast damage data that are more appropriate to use for gas and vapour explosions. Useful data on the overpressures required to damage brick walls, light partition walls, glass windows, etc were obtained from experimental work to study the effect of gas explosions within buildings and enclosures [7]. Further information on the response of buildings and building components can be found in References 8 to 11. The data is often presented in the form of damage-probability-loading diagrams [10]. Glass windows and many other building components show a wide variation in the overpressure and impulse required for damage, due to such factors as age, existing defects and quality of construction.

The effect of an external explosion on process plant is extremely complex. Process plant items are generally ductile targets and hence impulse is as important a parameter as peak side-on overpressure. For definitive solutions to a particular problem it is recommended that specialist advice is sought. A general, very coarse, indication of the probable amount of damage may be obtained from Figure 5.1 [12], with data obtained from nuclear bomb explosions, which have a longer positive phase duration than a typical vapour cloud explosion. The horizontal lines represent the range of 1 to 99 % probability of failure. It does, however, need to be stressed that the output is very coarse and specialist advice should be sought before reaching any final conclusion.

5.3 EFFECTS ON PEOPLE

PRIMARY EFFECTS

The principal fatality (as opposed to injury) mechanism is that of lung damage. Survivability curves for lung damage, in terms of scaled overpressure versus scaled impulse, are given in Figure 5.2 (see page 56). It can be seen that significant injury is caused at fairly high blast overpressure. A worked example demonstrating the use of this graph is given in Chapter 6.

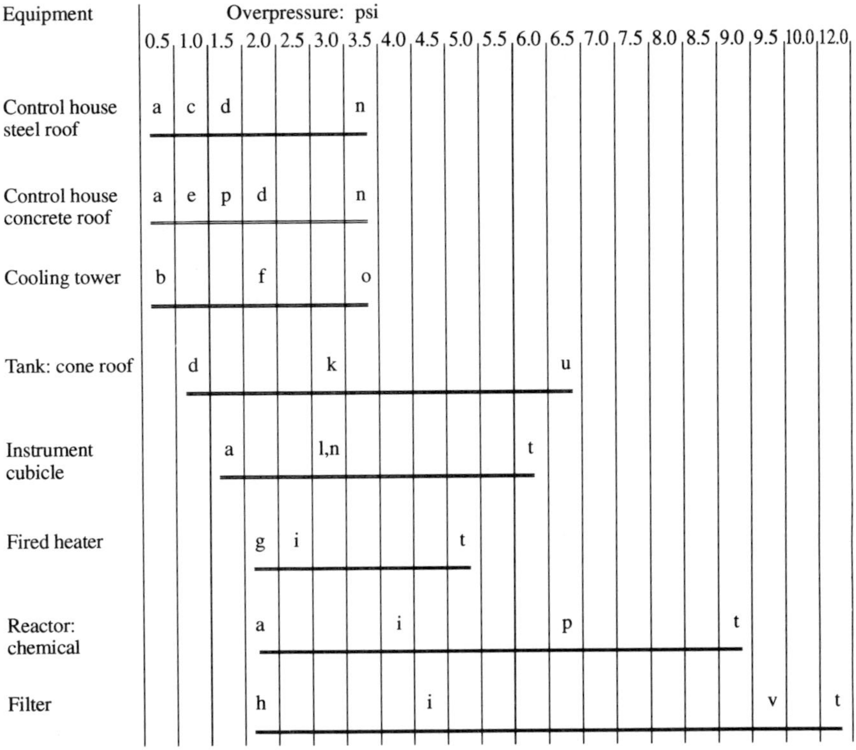

Figure 5.1 Damage to plant (continued on pages 54 and 55). (Reproduced by permission of US Department of Interior, Office of Oil and Gas.)

Code

a: windows and guages break
b: louves fail at 0.3–0.5 psi
c: switchgear is damaged from roof collapse
d: roof collapse
e: instruments are damaged
f: inner parts are damaged
g: bricks crack
h: debris missile damage occurs
i: unit moves and pipe break
j: bracing fails
k: unit uplifts (half filled)
l: power lines are severed
m: controls are damaged
n: block walls fail
o: frame collapses
p: frame deforms
q: case is damaged
r: frame cracks
s: piping breaks
t: unit overturns or is destroyed
u: unit uplifts (0.9 filled)
v: unit moves on foundation

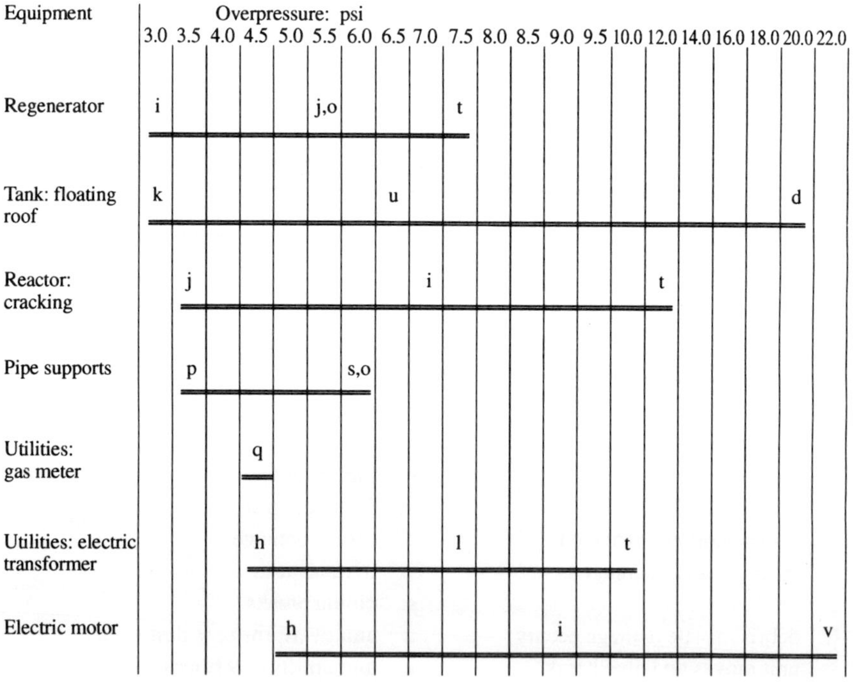

Figure 5.1 (Continued)

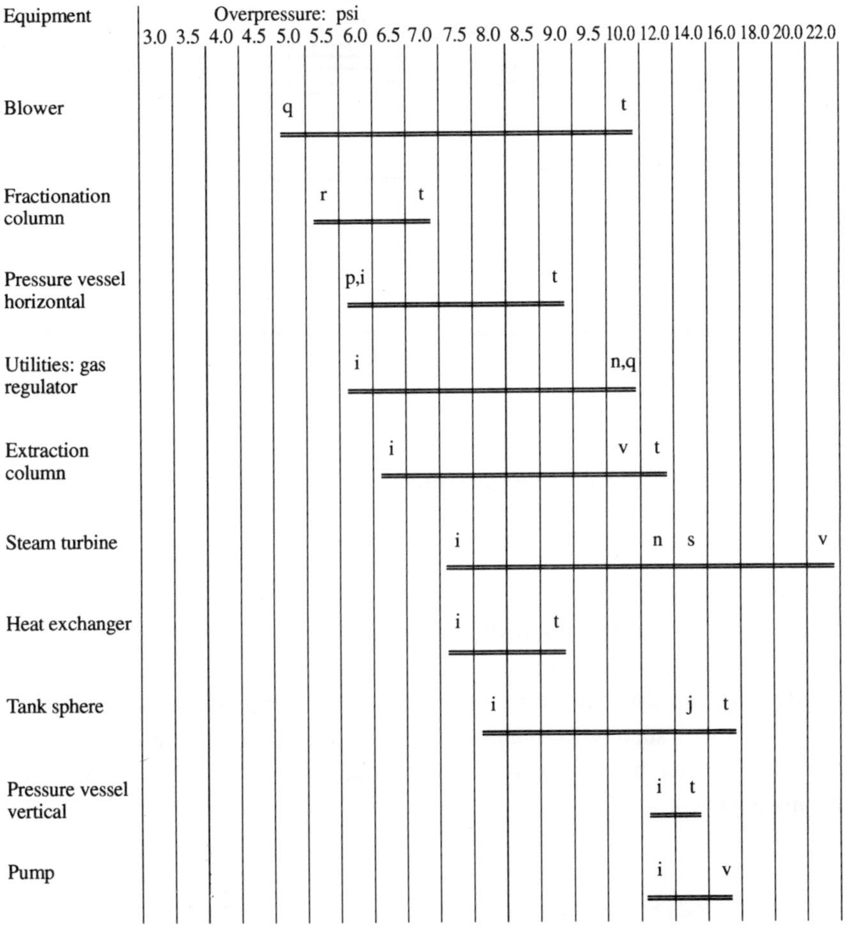

Figure 5.1 (Continued)

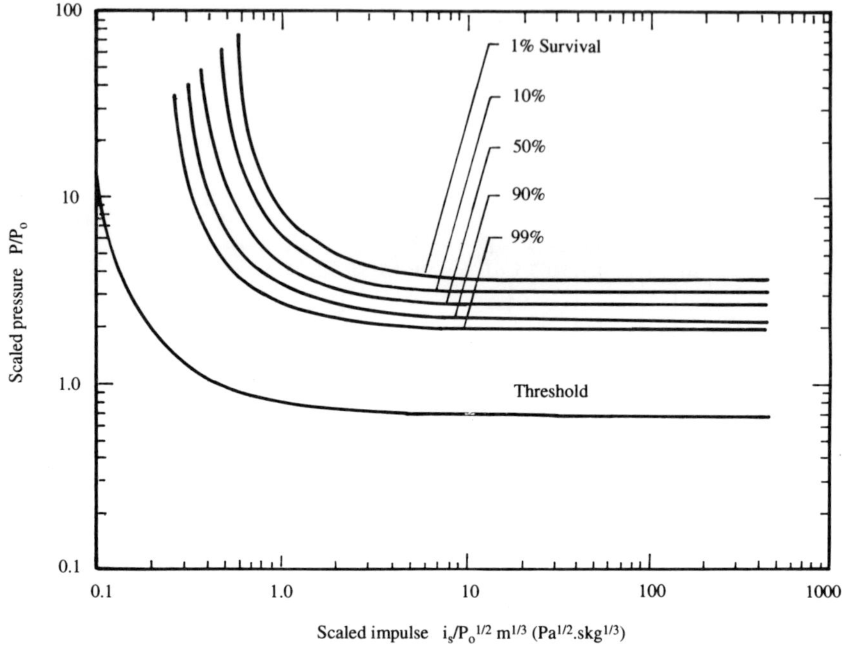

Figure 5.2 Survivability from primary blast effects. P_s = ambient pressure, i_s = impulse per square metre of the blast waves, P = overpressure, m = mass of target person. (Baker *et al* [2].)

SECONDARY EFFECTS

Considerable damage to ordinary housing occurs at very low blast overpressures (2–3 psi). This can result in impact from collapsing structural components and also to the projection of fragments from materials such as glass. Associated with such serious damage and collapse of buildings are injuries and fatalities. A number of explosion consequence models are available for estimating the probability of injury/fatality (for examples see References 13, 14 and 15).

These models are based upon analysis of the effects of actual explosions, and generally predict relationships between fatality probability and scaled distance. Whilst this is an acceptable simplification for predictive purposes, there are other parameters (topography, shielding, etc) which can

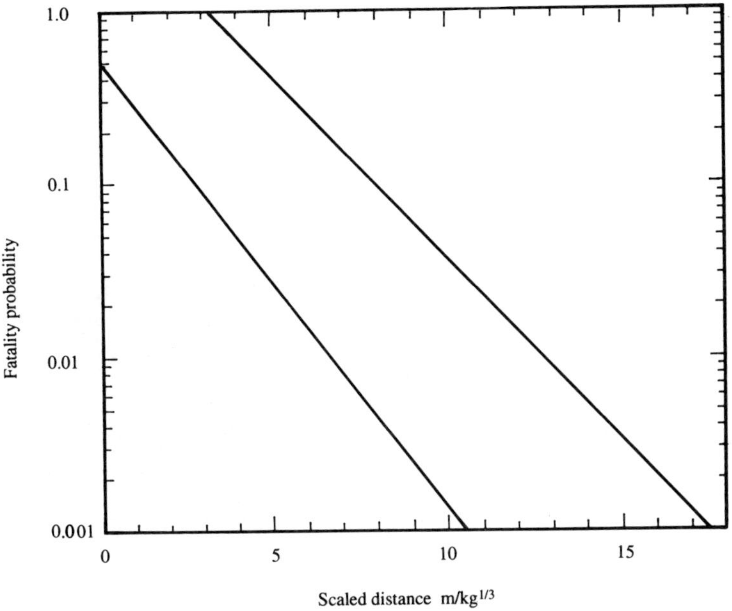

Figure 5.3 Survivability from secondary effects of blast. Distance scaled by the cube root of the mass of the TNT equivalent of the explosions.

affect the outcome and it is not surprising that the various models do not provide exactly similar predictions in all cases. However, the predictions generally fall into the band shown in Figure 5.3.

TERTIARY EFFECTS

The principal mode of death/injury is that of whole body translation that is, impacts from being 'blown over' though the effects of differential organ movement can also occur at very high levels of overpressure. Significant levels of overpressure are required for occurrence of body translation. For situations where people may be blown against walls/buildings, fatality probabilities may be estimated from plots of peak side-on overpressure versus scaled impulse as given in Figure 5.4 on page 58. A worked example demonstrating the use of this graph is given in Chapter 6.

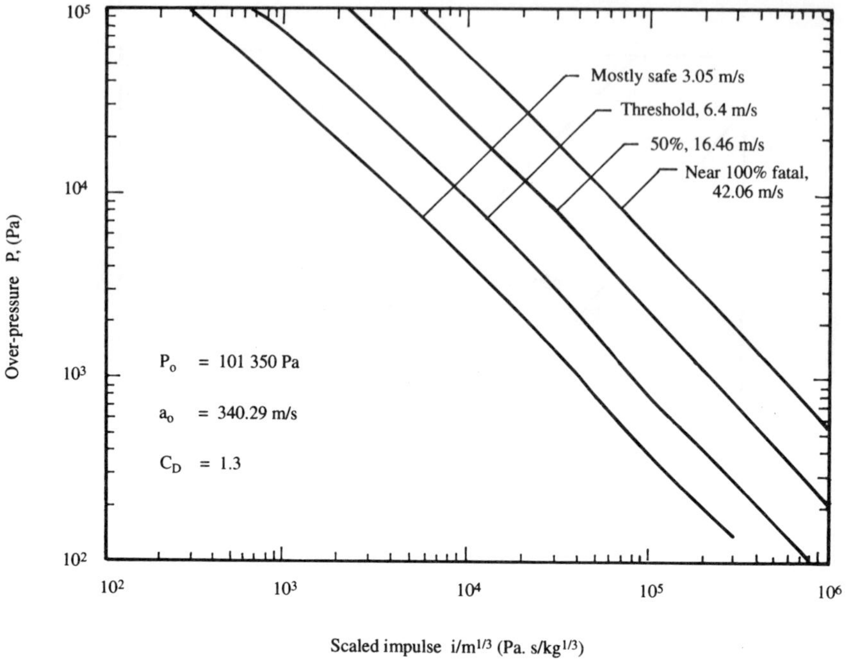

Figure 5.4 Survivability from tertiary effects of blast. P = overpressure, m = mass of target person, showing also velocity of body translation. (Baker *et al* [2].) (Reproduced with permission of Elsevier, Amsterdam, The Netherlands.)

5.4 RESPONSE OF STRUCTURES TO MISSILES

Missiles could, in principle, cause two types of damage — local and global. Local damage is governed by a set of conditions, referred to as terminal ballistics. They relate to the interaction between a 'target' and a projectile impacting the target at a velocity sufficient to cause local dynamic loading.

Missiles have historically been categorized as soft, semi-hard or hard depending on the degree of missile deformation and the coupling between target deformation and impact loading.

Soft missiles experience significant deformation during impact — for example, buckling in the case of a thin-walled tube which may crush by several centimetres while the target deforms only a few millimetres — and the structural response to the impact is generally assessed using a time-dependent load applied to the impact zone. The load can be estimated relatively simply by considering only the deformation of the missile and the problem is generally one of global response or structural collapse rather than local perforation.

Semi-hard missiles experience some deformation during impact and there is some coupling between target deformation and impact load — for example, in the case of an end-cap missile which deforms by a few millimetres when impacting a target which deforms by similar amount. Local response in terms of target perforation may be dealt with by correlation. Because of the reduced deformations in the missile, estimating the impact loads and target response by hand calculation is generally difficult. Techniques such as finite element calculations using sophisticated computer codes for global response assessments are needed.

Hard missiles experience negligible deformation during impact and there is some coupling determined only by the target response. An example of a hard missile would be a turbine blade impacting a thin target panel end-on. Local response in terms of target perforation can be dealt with by empirical correlations based on test data. Global response assessment can be made using simplified mechanistic models of the target response or, in the case of complex geometries, by finite element analysis.

Empirical correlations for use with hard missiles are often based on idealized cylindrical missiles with flat ends. Shape factors are defined for various nose geometries such as ogives, hemispheres and cones. For missiles with non–circular cross sections, the diameter, d_e, of an equivalent cylindrical missile can be defined on the basis of the perimeter, p, of the shadow the missile creates on the target by:

$$d_e = p / \pi$$

This equivalence has been observed for missiles with rectangular, square and triangular and complex sections. An equivalence of this nature is expected where the ultimate failure mode of the targets is by localized shearing around the missile edges. What is less expected is that the equivalence has been observed even where the actual amount of energy involved in this final process is small in comparison with that absorbed in membrane extension or by bending.

LOCAL RESPONSE/DAMAGE

The simplest formula for predicting the perforation energy, E, of flat steel plates is that derived by the Ballistics Research Laboratory (BRL) and is given by the following equation:

$$E = 1.44 \times 10^9 \, (td)^{1.5}$$

where t is the target thickness and d is the missile diameter.

Neilson [16] applied the above formula to a series of experimental results for projectiles weighing between 1.0 and 19.8 kg impacting steel targets of 1 to 25 mm thickness at velocities of 10 to 100 m/sec. There was a large scatter of results but the bulk of results were predicted to 30%.

Neilson [17] carried out tests on 150 mm nominal bore pipes with wall thicknesses between 7 and 18 mm. Missile diameters and masses ranged from 25 to 170 mm, and 4 to 50 kg respectively. The perforation energies of solid billets were correlated by the equation:

$$E/(ud^3) = A \, (t / d)^{1.7} \, (d / D)^{0.5}$$

where:

D = diameter of the target pipe;
t = pipe wall thickness;
d = missile diameter;
u = ultimate tensile strength of target material.

Methods to calculate local indentation of pipes which are not perforated by impactors are proposed by Ellinas [18]. The depth of the indentation, d, for a point loading is related to the impact energy, E, and pipe diameter, D, by the equation:

$$E = 100 \, M_o^{0.5} \, (d \sigma / D)$$

where

M_o = $0.25 \, t^2 s$;
t = wall thickness;
σ = yield stress.

GLOBAL RESPONSE/DAMAGE

The global response of structures to impact is a function of the received impact load imposed by the missile. For soft missiles, a simple procedure is proposed in Reference 2. The method is for determining the load imposed on a target by a collapsing missile. The procedure is given by the following equations:

$$F(t) = P_d(X,V) + U(x) \, V^2$$

$$A\ (X,V) = P_d\ (X,V)\ /\int_X^L U\ (k)\ dk$$

where:

$F(t)$	=	the instantaneous force exerted by a material of the missile at a distance X from the impact zone;
$Pd(X,V)$	=	is the dynamic crushing strength of the missile at position X and crushing velocity V;
$U(X)$	=	the mass per unit length at X;
V	=	the instantaneous velocity of the uncrushed portion of missile;
$A(X,V)$	=	the deceleration force in the uncrushed portion;
$U(k)$	=	the mass per unit length of the crushed part of the missile;
L	=	length of the missile.

The solution of the above equations can be accomplished by a numerical time integration scheme [19]. For hard or semi-hard missiles the interaction between target and missile prevents a simple expression for contact force being derived. However, Langseth and Larsen [20] describe an empirical model for estimating contact loads in low velocity impact for a flat faced projectile. The force, F, at failure is given by:

$$F = \pi\ d\ t\ \tau c$$

$$\tau c = \sigma\ [\ 0.4\ (t/d) + 0.42\ ]$$

where

d	=	missile diameter;
t	=	target thickness;
s	=	static ultimate tensile strength.

REFERENCES IN CHAPTER 5

1. Glasstone, S. and Dolan, P.J., 1977, *The Effects of Nuclear Weapons*, Third edition (US Department of Defence).
2. Baker, W.E., Cox, P.A., Westine, P.S., Kulesz, J.J. and Strehlow, R.A., 1983, *Explosion Hazards and Evaluation* (Elsevier, Amsterdam, The Netherlands).
3. The prediction of the pressure loading on structures resulting from an explosion, *Offshore Technology Report OTI 92 594*, 1993 (HMSO, UK).
4. Hoerner, S.F., 1958, *Fluid Dynamic Drag* (148 Busteed, Midland Park, NJ, USA).

5. Uppard, J.E., 1967, A survey of reports on experimental studies of forces exerted on model structures by the wind and by blast waves, *AWRE Report No E 6/67* (Atomic Weapons Research Establishment, UKAEA).
6. Scilly, N.F. and High, W.G., 1986, *5th Intl Symposium on Loss Prevention and Safety Promotion in the Process Industries.*
7. Harris, R.J., 1983, *The Investigation and Control of Gas Explosions in Buildings and Heating Plant* (E. and F. N. Spon, London, UK).
8. Pritchard, D.K., 1989, A review of methods for predicting blast damage from vapour cloud explosions, *Journal of Loss Prevention in the Process Industries,* 2: 187–193.
9. Mainstone, R.J., 1976, The response of buildings to accidental explosions, *BRE Current Paper CP 24/76*, (Building Research Establishment, UK).
10. Zeeuwen, J.P. and Schippers, J., 1980, Simplified blast hazard analysis, *Report No PML 1980–130* (Prins Maurits Laboratorium, TNO, The Netherlands).
11. Mercx, W.P.M., Weerheijm, J. and Verhagen, Th.L.A., 1991, Some considerations on the damage criteria and safety distances for industrial explosions, *IChemE Symposium Series No 124, Hazards XI: New Directions in Process Safety* (IChemE, UK).
12. Stephens, M.M., February 1970, *Minimising Damage to Refineries* (US Department of the Interior, Office of Oil and Gas).
13. *Canvey: Summary of an Investigation of Potential Hazards from Operations in the Canvey Island Thurrock Area*, 1978 (HMSO, UK).
14. *Canvey: A Review of Potential Hazards from Operations in the Canvey Island/ Thurrock Area Three Years after Publication of the Canvey Report*, 1981 (HMSO, UK)
15. Health and Safety Commission Report and Appendices, *Advisory Committee on Dangerous Substances; major aspects of the transport of dangerous substances* (HMSO, UK).
16. Neilson, A.J., 1985, Empirical equations for the perforation of mild steel plates, *Int J Impact Engng*, 3 (2): 137–142.
17. Neilson, A.J., 1987, Impact resistance of mild steel pipes: an experimental study, *AEEW–R2125.*
18. Ellinas, C.P. and Walker, A.C., 1983, Effects of damage on offshore tubular bracing members, *IABSE Colloquium, Ship Collision with Bridges and Offshore Structures, Copenhagen.*
19. Barr, P., 1990, *Guidelines for the Design and Assessment of Concrete Structures Subjected to Impact* (AEA–SRD).
20. Langseth, M. and Larsen, P.K., 1990, Dropped objects plugging capacity of steel plates: an experimental investigation, *Int J Impact Engng,* 3 (9): 289–316.

6 EXAMPLE CALCULATIONS

6.1 VAPOUR CLOUD EXPLOSION OVERPRESSURES

EXAMPLE PROBLEM

An incident results in a release of natural gas*. The mass of natural gas within the flammable envelope has been estimated at 1200 kg. The release is into a highly congested region of plant which has overall dimensions 30 m width, 30 m length and 8 m height (volume = 7200 m^3). If the vapour cloud is ignited, at what distance from the edge of the congestion will a peak incident overpressure of 0.004 bar be registered?

The results of the calculations are given to the nearest metre, although due to the approximate nature of the methods, and the lack of accuracy in the estimate of cloud size, such accuracy cannot be justified. In the extreme far field, as in this example, the calculations show a wide range of estimated distance at the 0.004 bar overpressure level. Should the calculations be repeated at a higher level of overpressure, the differences are reduced.

THE TNT EQUIVALENCE METHOD

From the total mass of fuel (hydrocarbon) in the release (1200 kg) determine;

$$\text{TNT equivalence} = \text{energy ratio factor} \times \text{efficiency factor}$$

$$= 10 \times 0.04 = 0.4$$

$$\text{Equivalent mass of TNT} = 1200 \times 0.4 = 480 \text{ kg}$$

From Figure 3.2 (see page 18) an incident peak overpressure of 0.04 bar occurs at a scaled radius of 25, hence the actual radius (r) from the explosion centre is

---

* Natural gas refers to a mixture of hydrocarbon gases but is predominantly methane (~95%). Ethane is the other dominant component.

given by:

$$r = 25 \times (480)^{1/3} = 195.7 \text{ m}$$

Without more explicit information on the shape of the cloud the equivalent explosion centre — that is, the position of the TNT charge — can be assumed to be at the cloud edge from which it follows that 0.04 bar peak overpressure occurs at approximately 196 m from the cloud edge.

BRITISH GAS VCE METHOD

First calculate the volume V (m^3), defined as the smaller of the congested region (V_1) and that (V_2) of a cloud of stoichiometric concentration containing 1200 kg of natural gas.

$$V_1 = 7200 \text{ m}^3$$

$$V_2 = 1200 / 0.08 = 15{,}000 \text{ m}^3$$

It follows that $V = 7200$ m^3 and the equivalent mass of TNT is therefore:

$$M_{TNT} = 16 \times 7200 \times 0.01 = 1152 \text{ kg}$$

The scaled distance corresponding to a 0.04 bar overpressure is 25 (see Figure 3.2, page 18) thus:

$$sd = 25 = 1.6 = r/M_{TNT}^{1/3}$$

hence:

$$r = 23.4 \times 10.48 = 245.2 \text{ m}$$

from which it follows that an incident 0.04 bar peak overpressure is registered a distance of 245 m from the edge of the congested region.

TNO MULTI-ENERGY METHOD

Step 1

V_1 is defined as the total volume within any group of obstacles, thus

$$V_1 = 30 \times 30 \times 8 = 7200 \text{ m}^3$$

V_2 is the largest cloud the stored inventory is capable of producing (assuming a stoichiometric mixture), hence:

$$V_2 = 1200 / 0.08 = 15{,}000 \text{ m}^3$$

Since V_1 is less than V_2 then:

$$V_1 = 7200 \text{ m}^3 \text{ and } V_3 = 7800 \text{ m}^3$$

Step 2
Calculate the energy in volume V_1:

$$E_1 = 3.5 \times 7200 = 25{,}200 \text{ MJ}$$

Calculate the energy in volume V_3:

$$E_3 = 3.5 \times 7800 = 27{,}300 \text{ MJ}$$

Step 3
For the congested volume V_1, using curve 7 of Figure 4.1 on page 33 (the curve which starts at 1 bar overpressure) an incident peak overpressure of 0.04 bar occurs at a scaled radius of approximately $R = 6$.

For the open volume V_3, using curve 2 of Figure 4.1 (the curve which starts at 0.02 bar overpressure) the overpressure will be below 0.04 bar everywhere.

Step 4
The actual radius (r) from the explosion centre (assumed to be at the centre of a hemispherical cloud which is itself centred on the congested region) is therefore given by:

$$r = 6 \times (25.2 \times 10^9/10^5)^{1/3} = 6 \times 63.16 = 379.0 \text{ m}$$

An incident peak overpressure of 0.04 bar is therefore realised at a distance of 364 m (379 – 30/2) from the edge of the congested region.

SHELL METHOD

Step 1

The volume (V_1) within 2 m from any group of obstacles is given by:

$$V_1 = 34 \times 34 \times 10 = 11{,}560 \text{ m}^3$$

and this is doubled to give:

$$V_1 = 23{,}120 \text{ m}^3$$

The total volume of the vapour cloud, V_2, is:

$$V_2 = 15{,}000 \text{ m}^3$$

hence, since V_2 is less than V_1, the value of V_1 is replaced by V_2:

$$V_1 = 15{,}000 \text{ m}^3$$

The volume V_1 can be contained within a hemisphere of radius R_1, where:

$$R_1 = (3 \times 15000)/(2 \times p) = 19.2 \text{ m}$$

Step 2

Referring to the decision tree in Figure 4.4 (see page 37), the region is obstructed but not confined, natural gas is generally regarded as being less reactive than ethylene, and the high congestion in V_1 justifies a reference overpressure of 1 bar, ($P_{ref} = 1$).

Step 3

For volume V_1 the distance ($R : m$) from R_1 at which incident overpressure, P_{max}, of 0.04 bar is realized is therefore given by:

$$R = R_1 \times (P_{max} / P_{ref} - 1)$$

$$R = 19.2 \times (1.0/0.04 - 1) = 460.8 \text{ m}$$

Assuming the equivalent hemispherical vapour cloud is central to the congested region it follows that 0.04 bar incident overpressure is realized at a distance of 465 m, (460.8 + 19.2 – 30/2) from the edge of the congested region.

Volume V_3 is zero, so calculation of overpressure for an unconfined region is not relevant.

6.2 INJURY TO PEOPLE

What would be the chance of survival of a 75 kg man, at atmospheric pressure $P_a = 103$ kPa (14.75 psi), subject to a blast overpressure P_s of 310 kPa (44.4 psig) and a duration t, of 10 ms?

PRIMARY EFFECTS (LUNG DAMAGE)

Scaled overpressure $\overline{P}_s = P_s / P_a = 310/103 = 3.01$

$$\text{Primary effect scaled impulse } \bar{i}_s = \frac{i_s}{P_a^{0.5} \times m^{0.33}}$$

$$\text{where } \bar{i}_s = \text{(approx)} \ \frac{P_s \times t}{2} = 310{,}000 \times 0.01 \times 0.5 = 1550 \text{ Pa s}$$

$$i_s = \frac{1550}{103{,}000^{0.5} \times 75^{0.33}} = 1.15\ (Pa^{0.5}\ \text{s/kg}^{0.33})$$

From Figure 5.2 (see page 56) we can see that this combination of scaled overpressure versus scaled impulse gives a 90% chance of survival.

SURVIVABILITY VIA WHOLE BODY TRANSLATION

Incident overpressure $P_s = 310{,}000$ Pa

Tertiary effect scaled impulse:

$$\bar{i}_s = i/m_s\, 0.33 = 1550/75^{0.33} = 368\ (\text{Pa } s/\text{kg}^{0.33})$$

From Figure 5.4 (see page 58) we see that this combination of incident overpressure versus scaled impulse is slightly off the graph. Extrapolation of the survivability curves suggests a survivability probability approximately half way between the threshold and 50–75 %.

Thus the man has a good prospect of survival from the primary effects of overpressure, but a lower prospect of survival from impact after being blown over.

7. GLOSSARY

Blast wave
A pressure pulse formed by an explosion.

Confined explosion
An explosion inside a closed system (for example, a vessel or building).

Deflagration
The chemical reaction of a substance in which the reaction front advances into the unreacted substance at less than sonic velocity. Where a blast wave is produced which has the potential to cause damage, the term explosive deflagration is usually used.

Distance (non-dimensional)
A distance combined with other parameters such that the group of parameters including distance does not have any unit of dimension.

Distance (scaled)
A distance combined with other parameter such that the combination has a unit of dimension. A non-dimensional distance is also a scaled distance, but a scaled distance might not be non-dimensional.

Detonation
An explosion caused by the extremely rapid chemical reaction of a substance in which the reaction front advances into the unreacted substance at greater than sonic velocity.

Diffraction loading
Caused by the bending (diffraction) of a blast wave around a structure engulfing it and causing crushing forces. Damage caused is related to the overpressure and the relationship between structure size or length (in the direction of the blast wave) and the blast wave duration.

Drag loading

Caused by the blast wind forces. Its magnitude is dependent on the size, shape and orientation of a structure and the magnitude and duration of the dynamic overpressure.

Duration

The time taken for the pressure pulse to decline to zero is known as the positive phase duration, usually shortened to duration.

Dynamic overpressure

The pressure generated by the wind travelling behind the blast wave.

Epicentre

The ground location beneath the inferred centre of a vapour explosion.

Explosion

A release of energy which causes a pressure discontinuity or blast wave.

Explosion 'efficiency'

The ratio of the energy in the blast wave to the energy theoretically available from the heat of combustion, usually expressed as a percentage.

Explosive power

The amount of explosion energy released relative to the explosion energy released by the same mass of TNT, expressed as a percentage.

$$\text{Explosive yield} = \text{weight} \times \text{TNT equivalence} \times \text{efficiency}$$

$$\text{Explosive power} \times \text{weight} = \frac{\text{Explosive power} \times \text{weight}}{100}$$

Far field

The region at the outer limits of the explosion where the overpressure decay is similar to that of a TNT explosion.

Missiles

Fragments or whole systems which are projected by a release of energy.

Overpressure

For a pressure pulse (blast wave), the pressure developed above atmospheric pressure at any stage or location is called the overpressure. Overpressure is sometimes used to describe exposure of equipment to pressures in excess of the design pressure, but the term overpressurization is preferred for this purpose.

Peak positive overpressure

The maximum overpressure generated at a specified distance is called the peak positive overpressure.

Pressure blast

The rupture of a system under pressure, resulting in the formation of a blast wave and missiles which may have the potential to cause damage.

Rapid phase transition

The rapid change of state of a substance which may produce a blast wave and missiles.

Reflected overpressure

If a 'rigid' object was perpendicular to the advance of the blast wave (that is, facing), the object would reflect and diffract the wave. Due to this reflection, the object will experience an effective overpressure, usually called the reflected overpressure. It may vary from two to perhaps eight times the incident overpressure depending on the value of the incident overpressure.

Scaled distance

See 'Distance (scaled)'.

Shockwave

A pressure pulse formed by an explosion in which a sharp discontinuity in pressure is created as the wave travels through a fluid medium at greater than sonic velocity.

Side-on overpressure

If a pressure-sensitive device which offered no obstruction to the passage of the blast wave was placed in its path (that is, one which was facing sideways in relation to its advance), the device would record side-on overpressure.

TNT equivalent
The amount of TNT (trinitrotoluene) which would produce the same damage effects as those of the explosion under consideration. For non-dense phase explosions the equivalence has meaning only at a considerable distance where the nature of the blast wave arising is comparable with that of TNT.

Vapour cloud explosion (VCE)
The preferred term for an explosion in the open air of a cloud made up of a mixture of a flammable vapour or gas with air.

INDEX